AS

VISUAL REVISION GUIDE

Learning Servi...   Please ... on or be...

SUCCESS

BIOLOGY

Byron Dawson

Contents

Biological molecules

Cells

Exchange and transport

Reproduction

Chromosomes, genes and DNA

Human health and disease

Ecology and the environment

Carbohydrates

How carbohydrates are formed

- All carbohydrates are made up from three elements:

  CARBON   HYDROGEN   OXYGEN $\longrightarrow$

  The basic formula is always $(CH_2O)n$.

- This is why they are called <u>carbohydrates</u>.
- The 'n' represents the number of times the unit is repeated.

Monosaccharides

The simplest forms of carbohydrates are called **monosaccharides**.
- They are sometimes called simple sugars.
- Glucose and fructose are two monosaccharides.
- The formula for glucose is $C_6H_{12}O_6$.
- Glucose can exist in two forms, α and β. One is a mirror image of the other.

Look at the structural formula for glucose. Count the number of each kind of atom and see if it matches the basic formula $C_6H_{12}O_6$.

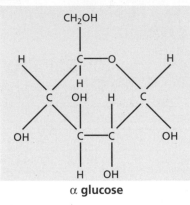

α glucose

Although glucose has 24 atoms, it is a very small molecule. This is why it can pass through the gut wall and into the bloodstream. Glucose molecules form the building blocks for much bigger molecules.

Disaccharides

- A single glucose molecule is called a **monomer**.
- Two monomers can join to form a **disaccharide**.
- When two monomers join, a molecule of water is released.
- This is called a **condensation reaction**.
- The link that holds the two glucose molecules together is called a **glycosidic bond**.
- The disaccharide produced is a sugar called **maltose**.
- Sucrose is formed in the same way when glucose and fructose join together.

  $C_6H_{12}O_6 + C_6H_{12}O_6 = C_{12}H_{22}O_{11} + H_2O$

- If a molecule of water is added, the bond can be broken to give two molecules of six carbon sugars. This is called **hydrolysis**.

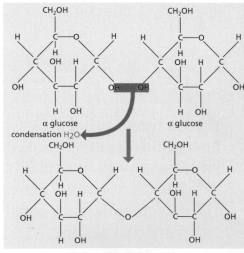

Maltose

Polysaccharides

- **Polysaccharides** are produced when many glucose molecules are joined together.
- When lots of monomers join together it is called a **polymer**.
- Starch is a polymer produced when lots of glucose monomers join together with glycosidic bonds.

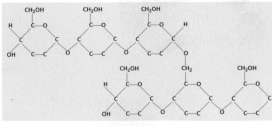

Part of a branched section of a starch molecule

The diagram shows a type of starch molecule called **amylopectin**. It has a branched side chain.

Hydrolysis of polysaccharides

Polysaccharides can be broken down by the addition of water. However, this may take a long time. Enzymes can speed up the process.

Try this ...

- Bread contains starch. Saliva contains the enzyme **amylase**. Chew a piece of bread and leave it in your cheek for about 15 minutes. It will start to taste sweet as the polysaccharide is split up into sugars.

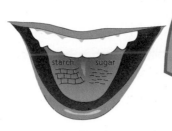

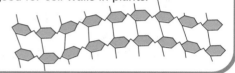

EXAMINER'S TOP TIP
Hydro means 'water'; lysis means 'to break down'.

Other polysaccharides

Glycogen is made of long branched chains of α glucose molecules. It is more soluble than starch and can be stored in the liver as an energy reserve.

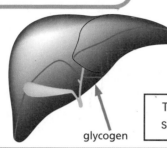

glycogen

Cellulose

Cellulose is formed from long chains of β glucose. The chains form a lattice that gives cellulose its strength and toughness. These characteristics make it good for cell walls in plants.

This is called relating structure to function.

Which sugars make which polysaccharide?

Polysaccharide	Disaccharide	Monosaccharide
starch	maltose	glucose
glycogen	sucrose	fructose
cellulose	lactose	galactose

Pectin

- Pectins are polysaccharides that are bound together with **calcium pectate**.
- They are found in plants.
- They help plant cells join together.
- They help jam to set.
- They can make wines cloudy.

How sweet is sugar?

Some sugars taste sweeter than others. The table shows the sweetness of different sugars, as compared with sucrose – the ordinary white sugar used in cooking.

- Lactose 0.2
- Maltose 0.3
- Galactose 0.3
- Glucose 0.7
- Sucrose 1.0
- Fructose 1.7

Quick test

1 What do we call sugars that contain six atoms of carbon?
2 What is released when two molecules of glucose join together?
3 What is the name of the reaction when two molecules of glucose join together?
4 What carbohydrate is produced when two molecules of glucose join together?
5 What is the name of the bond that joins two molecules of glucose together?
6 What do we call two monosaccharide sugars joined together?
7 Name two other important polysaccharides.
8 What is the name of the reaction when polysaccharides are split into monosaccharides?

1. monosaccharides and hexoses 2. water 3. condensation reaction 4. maltose 5. glycosidic 6. disaccharide 7. glycogen, cellulose 8. hydrolysis

5

Lipids

What are lipids?

Lipids contain exactly the same elements as carbohydrates:

CARBON

HYDROGEN

OXYGEN

But their appearance is quite different to carbohydrates. They are fatty, oily or waxy.

- Carbohydrates always contain twice as much hydrogen as oxygen.
- Lipids always contain proportionately less oxygen.
- Thus they can release far more energy when they are oxidised.
- This makes them useful for storing energy in living things.
- Because they do not dissolve in water, lipids stay put when stored. This makes them even more useful.
- They are also used to insulate living things from the cold.
- They provide electrical insulation for neurones.
- They form part of the cell membrane.

How are fats formed?

- Fats are formed when **three fatty acids** react with **glycerol**.
- This is why they are often called **triglycerides**.

$$
\begin{array}{l}
O \\
\| \\
R - C - OH \\
\text{fatty acids}
\end{array}
\qquad
\begin{array}{l}
H \\
| \\
H-C-O-H \\
H-C-O-H \\
H-C-O-H \\
| \\
H \\
\text{glycerol}
\end{array}
$$

R = groups such as CH_3 or CH_2OH

- Notice that water is given off by the reaction. This is another example of a **condensation reaction**.
- Another similarity with carbohydrates is that when an enzyme and water are added the reaction is reversed. As with carbohydrates, this is called **hydrolysis**.

- The bond that joins the fatty acids and glycerol is called an **ester bond**.

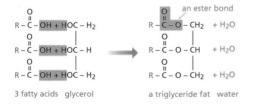

3 fatty acids   glycerol   a triglyceride fat   water

Saturated fats

- **Saturated fats** have no C=C (double bonds).
- This is because they are formed from saturated fatty acids which also have no C=C bonds.
- Stearic acid is a **saturated fatty acid**. When combined with glycerol it forms saturated fats.

$$H-C-C-C-C-C-C-C-C-C-C-C-C-C-C-C-C-C-C-OH$$

A saturated fatty acid

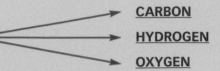

EXAMINER'S TOP TIP
How do you remember which reaction is which? Think about condensation on a cold surface – that is when water appears.

Unsaturated fats

$$H-C-C-C-C-C-C-C-C-C=C-C-C-C-C-C-C-C-C-OH$$

An unsaturated fatty acid

- **Unsaturated fats** have a C=C (double bond).
- They are formed from unsaturated fatty acids.
- Oleic acid is an **unsaturated fatty acid** which can form unsaturated fats.

Phospholipids

Lipids play an important role in cell membranes. But first they have to be converted into **phospholipids**. This is quite a simple process.

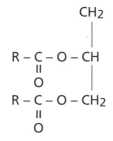

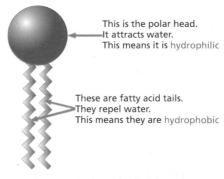

Lipids

- It requires a phosphate group

Phospholipids are often shown using this simplified diagram.

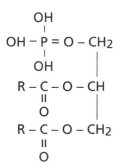

This is the polar head.
It attracts water.
This means it is hydrophilic

These are fatty acid tails.
They repel water.
This means they are hydrophobic

- Then one of the fatty acids in the triglyceride fat is removed.

A phospholipid

They are **polar molecules** – they have two distinct and different ends.

- It is replaced with the phosphate group to make a phospholipid.

Quick test

1. *How is the structure of lipids different from the structure of carbohydrates?*
2. *Which two molecules are used to produce a lipid?*
3. *What are lipids used for?*
4. *Why are fats called triglycerides?*
5. *What is the name of the bond that joins fatty acids to glycerol?*
6. *What is the difference between saturated and unsaturated fats?*
7. *How are phospholipids produced?*
8. *What does hydrophobic mean?*
9. *What does hydrophilic mean?*

1. Lipids contain proportionately less oxygen. 2. fatty acid and glycerol 3. energy storage, insulation of heat and electrical insulation of neurones 4. Three fatty acids combine with glycerol. 5. ester bond 6. Unsaturated fats have C=C double bonds. 7. A fatty acid in a triglyceride is replaced with a phosphate group. 8. repels water 9. attracts water

Proteins

How are proteins formed?

Proteins, just like carbohydrates and lipids, are also made from the elements

CARBON   HYDROGEN   OXYGEN

Unlike carbohydrates and lipids, proteins always contain NITROGEN.
Other elements such as Sulphur, Iron and Phosphorus are sometimes present – but not always.
HINT Proteins contain *CHON*

What are proteins made of?

- Proteins are made from building blocks called **amino acids**.
- All amino acids have the same basic structure.
- They all have an **amine group** and a **carboxylic group**.

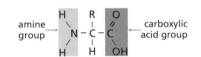

An amino acid

- But the R group can be different (such as CH_3 or C_2H_5) and this allows about 20 different amino acids.
- Think of the twenty amino acids as letters in an alphabet. Just as different combinations of letters make different words, different combinations of amino acids make different proteins.

- However, unlike words that contain only a few letters, proteins can contain hundreds of amino acids.

> **EXAMINER'S TOP TIP**
> As human beings, we need about ten of these amino acids in our diet. We make the other ten amino acids in our bodies. You do not need to know all the amino acids. Just learn the basic structure.

How are proteins made?

- Making protein is yet another example of a **condensation reaction**. Water is produced.
- Amino acids are joined together by **peptide bonds**.

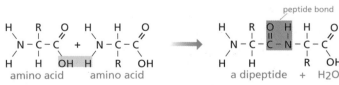

An amino acid

- Two amino acids joined together form a **dipeptide**.
- Lots of amino acids joined together form a **polypeptide**.
This linear sequence of amino acids is called the protein's **primary structure**.

peptide bond   amino acid

Secondary structures

Polypeptides can be twisted or coiled into shapes known as **secondary structures**. These structures give strength to proteins and form the special shapes of enzymes.

α-helix

- The α-helix is the most common secondary structure.
- The polypeptide chain is twisted into a helix, rather like a rope made of twisted strands.
- Weak hydrogen bonds link the carboxyl group on one turn with the amino group on the next turn.
- Keratin (the tough fibres found in hair and fingernails) is an example of a protein made from an α-helix.

amino acid

hydrogen bonds hold shape together

α-helix

β-pleated sheets

- β-pleated sheets are ribbon-shaped structures made from zigzag lines of amino acids.
- They are not as common as α-helix proteins.
- β-pleated proteins are very strong. Silk is one example.
- Lysozyme is a protein with a β-pleated protein. The strength enables the enzyme to maintain its shape.

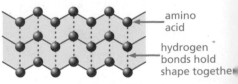
amino acid

hydrogen bonds hold shape together

β-pleated sheet

Fibrous proteins made from α-helixes

Some proteins are called **fibrous proteins**. They are made from several α-helix polypeptides twisted together, just like a rope made from twisted strands.

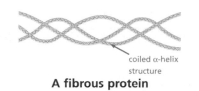

coiled α-helix structure

A fibrous protein

Tertiary structure

The name **tertiary structure** refers to the way the polypeptide is folded into a precise shape. The folded shape is held in a permanent position by various types of bonds, including:

- IONIC BONDS
- HYDROGEN BONDS
- DISULPHIDE BRIDGES

- The disulphide bridges are sulphur–sulphur bonds. They are very strong bonds.

Quaternary structure – globular proteins

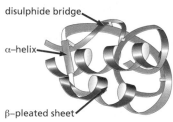

disulphide bridge

α–helix

β–pleated sheet

A quaternary structure

- They are often called **globular proteins**.
- Haemoglobin is a globular protein.
- It has four separate polypeptide chains.

- **Quaternary structure** is when two or more polypeptides are joined together in a complex folded shape.
- The polypeptides may be a mixture of α-helix and β-pleated sheets.

- In the picture, the α chains are shown in **brown** and the β chains are blue.
- The red discs are the **haem groups** which contain iron.
- The haem group is sometimes called a **prosthetic group**.
- The haem group helps to transport oxygen.

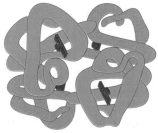

Haemoglobin

Quick test

1. What units are proteins made from?
2. What is produced when two of these units are joined together?
3. What is the name of the bond that joins these units together?
4. Describe the primary structure of a protein.
5. Name two kinds of secondary structures.
6. What kind of secondary structures are found in the enzyme lysozyme?
7. Explain the importance of secondary structures.
8. Name three kinds of bonds responsible for tertiary structure.
9. What is the difference between tertiary and quaternary protein structure?
10. What is a globular protein?
11. What is a fibrous protein?

1. amino acids 2. a dipeptide 3. a peptide bond 4. a linear chain of amino acids 5. α-helix and β-pleated sheet 6. Alpha helix and β-pleated sheet 7. They give strength to structural proteins and shape to enzymes. 8. ionic bonds, hydrogen bonds and disulphide bridges. 9. A tertiary structure has a folded polypeptide chain; a quaternary structure contains several polypeptide chains. 10. a quaternary structure with a complex folded shape 11. polypeptide strands of α-helix twisted together

Biochemical tests

Tests for carbohydrates

The following food tests can be used to identify carbohydrates, fats, and protein.

Benedict's test

This tests for reducing sugars. These are all monosaccharides, e.g. glucose and some disaccharides – but NOT sucrose.

- Add a few drops of **Benedict's solution** to the food solution.
- Heat in water bath until it boils.
- The presence of monosaccharides is indicated by an orange precipitate.

Non-reducing sugar test

This tests for sucrose.

- Carry out Benedict's test.
- If negative, attempt to hydrolyse the solution by heating with **dilute hydrochloric acid**. This breaks the glycosidic bond and releases the monomers.
- Neutralise with **dilute sodium hydroxide**.
- Perform Benedict's test for the second time. A positive result shows the presence of sucrose in the original solution.

Starch test

- Add a few drops of iodine in potassium iodide (**iodine solution)**.
- A blue-black colour indicates the presence of starch.

Emulsion test for lipids

- Add 2 cm³ of ethanol to the food solution in a test tube. Shake well.
- Add 2 cm³ of water to the solution in the test tube. Shake again.
- If the solution forms a milky white emulsion, this is a positive result, indicating the presence of fat.

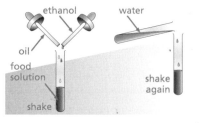

Biuret test for protein

- Add some **dilute copper sulphate** solution to the food solution in the test tube.
- Add some **dilute sodium hydroxide solution** to the test tube.
- If the solution gradually turns purple or lilac, this is a positive result, indicating the presence of protein.

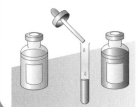

Separation of substances

Chromatography is a technique used to separate different substances from a sample mixture. It can be used to separate out the pigments in chlorophyll, or the different amino acids in a protein.

- A spot of an unknown mixture is placed on a sheet of chromatography paper and left to dry.
- A spot of a known mixture (for comparison) is also placed on the paper.
- The paper is suspended such that the paper touches the solvent but the dry spots do not.
- The solvent soaks up through the paper. The lighter molecules are carried more easily and quickly by the solvent.
- The mixture separates out into its components.
- The unknown components are then compared with the known ones.
- **Rf values** can then be calculated.

$$\text{Rf value} = \frac{\text{distance moved by substance}}{\text{distance moved by solvent front}}$$

- Every compound will have its own Rf value that never changes.
- Using different solvents allows different Rf values to be calculated for the same compound.

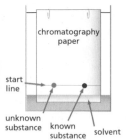

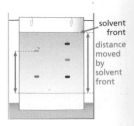

The importance of water to life

- 85% of a cell is water:
- Water is a very unusual material.
- Water's properties are unusual.

> **EXAMINER'S TOP TIP**
> Cohesion is also responsible for the surface tension on water, allowing animals like the water boatman to skate across the surface.

The unusual properties of water

1. Water has a high specific heat capacity
Its **high specific heat capacity** means that it can absorb a large amount of heat before its temperature starts to increase. It also means that it can lose a lot of heat before its temperature starts to decrease. This helps to keep the temperature inside cells fairly constant.

This also means that →

2. Water is a solvent
Water is an excellent **solvent**. All the chemical reactions in living organisms take place because the substances dissolve in water.

This is because →

3. Water molecules are cohesive
- Because water is a polar molecule, the – oxygen end of one molecule is attracted to the + hydrogen end of another water molecule.
- This means water molecules are attracted to each other by **hydrogen bonds**.
- This is called **cohesion**.
- Because the water molecules surround dissolved substances and stay together, water and dissolved substances are easily moved around living systems.

This means that ←

4. You can't squash water
When pressure on water is increased, it does not get smaller. This allows organisms like mammals to pump blood around their bodies under pressure.

5. Water has a high latent heat of vaporisation
Water absorbs a lot of heat to change state from a liquid to a gas – it has a **high latent heat of vaporisation**. This helps to keep us cool when we are hot. When sweat evaporates it cools us down and helps us maintain a constant body temperature.

6. Water is a polar molecule
The oxygen end of a water molecule has a – charge. The hydrogen end has a + charge. This makes it a **polar molecule**. The polar water molecules surround charged particles. This helps substances dissolve in them.

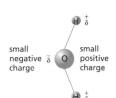

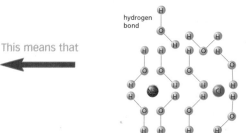

7. Water expands just before it freezes
Most substances get smaller when they get colder. When the temperature of water drops to 4 °C it starts to expand. This makes ice less dense so it floats. If ice did not float, then all the oceans of the world would be frozen apart from a few centimetres on the top. Life as we know it could not exist.

Quick test

1 Describe a test for reducing sugar.
2 Describe a test for non-reducing sugar.
3 Describe a test for lipids.
4 Describe a test for starch.

5 Describe a test for protein.
6 What is meant by Rf values?
7 Why is water called a polar molecule?
8 State three features of water that are important for living things.

1. Benedict's test – heat with Benedict's solution; orange-red is a positive result. 2. Perform benedict's test; hydrolyse with dilute hydrochloric acid; neutralise with dilute sodium hydroxide; Perform Benedict's test. 3. Emulsion test – add ethanol, water and shake; white emulsion is a positive result. 4. Add iodine solution. Turning black is a positive result. 5. Biuret test – add dilute copper sulphate and sodium hydroxide. Turning purple is a positive result. 6. distance substance moves / distance solvent front moves 7. The oxygen end is negative, hydrogen end is positive. 8. high specific heat capacity, high latent heat of vaporisation, good solvent, polar molecule, expands on freezing, cannot be squashed

Enzymes

What are enzymes?

Enzymes are sometimes called <u>biological catalysts</u>. A catalyst enables a reaction to take place rapidly. Enzymes are <u>globular proteins</u> and they have very specific shapes (see Proteins on p. 9).

(see Proteins on p. 9)

Lock and key model

The <u>**lock and key model**</u> can help us to understand how enzymes work.

Only one key will fit the lock.	→	Enzymes are very specific. One enzyme will do only one job.

A door can be locked or unlocked.	→	The reaction is reversible.

Keys can be used over and over again.	→	Enzymes can be used over and over again.

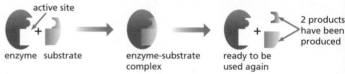

A catabolic reaction (substance broken down)

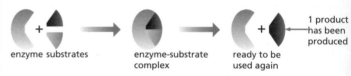

An anabolic reaction (substance used to build a new molecule)

How enzymes work

Reactions require an input of energy to get them started. This energy is called **activation energy**. Enzymes reduce the activation energy required to start a reaction, enabling the reaction to take place much more easily. Enzymes in our bodies lower the activation energy so that the reaction will take place at our body temperature of about 37 °C.

Progress of a reaction with and without enzyme

Think of a lump of coal. It contains lots of energy.

But it will not release this energy on its own.

First, we have to put a small amount of energy into the reaction to get it started – this is called the activation energy.

Enzyme concentration and the rate of reaction

A key can only open one door at a time. Even with lots of keys fitting doors with the same lock, there is a limit to how quickly the doors can be opened. It is the same with enzymes. Once all the active sites are in use, the reaction cannot go any faster.

The number of available enzyme molecules acts as a <u>**limiting factor**</u> to the rate of reaction.

Other factors affecting the rate of reactions

Temperature

- Enzymes have an **optimum temperature** at which they work best. Different enzymes in different organisms have different optimum temperatures.
- Below this temperature the molecules will have less kinetic energy. This means the molecules are moving more slowly. There will therefore be fewer collisions between the molecules and the reaction will be slower.
- Above this temperature the shape of the enzyme begins to change. This alters the shape of the active site (lock) so the substrate (key) can no longer fit and the reaction rate slows down. At very high temperatures, the change in shape may be permanent and the enzyme will no longer work.

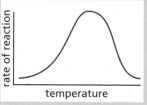

Effect of temperature on enzyme activity

pH

- Enzymes have an **optimum pH** at which they work best. Different enzymes in different organisms have different optimum pHs.
- pH can affect the bonds that form the secondary and tertiary structures of the enzyme. This in turn affects the shape of the enzyme.
- Any change in shape may alter the shape of the active site (lock) and the substrate (key) will no longer fit.
- The change in shape is not usually permanent. A return to the correct pH will usually mean the enzyme can work again.

Effect of pH on enzyme activity

EXAMINER'S TOP TIP
Always try to use the lock and key model to explain how enzymes work when you answer questions.

Competitive inhibition

Imagine someone trying to open a lock with the wrong key. The key gets jammed in the lock. The right key cannot be put in the lock while the wrong key is in the way.

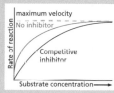

substrate
substrate cannot enter active site
substrate may now enter
competitive inhibitor
the inhibitor binds with the active site

An example of **competitive inhibition** is penicillin on bacteria.

Non-competitive inhibition

Imagine someone damages the keyhole on the lock. The keyhole is now the wrong shape and the key will no longer go into the lock.

non-competitive inhibitor
substrate
binding site
active site has changed
substrate has opportunity to enter

An example of **non-competitive inhibition** is the poison cyanide. All inhibitors are poisons.

The difference between competitive and non-competitive inhibition

In competitive inhibition, the reaction rate eventually reaches its maximum. It just takes longer.

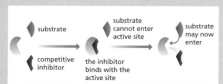

maximum velocity
No inhibitor
Competitive inhibitor

In non-competitive inhibition, the reaction rate never reaches its maximum.

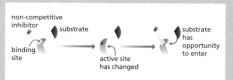

maximum velocity
No inhibitor
Non-competitive inhibitor

EXAMINER'S TOP TIP
Make sure you understand these two graphs.

Quick test

1 What is the name of one model used to help us understand how enzymes work?
2 Give three examples of how the model can help us understand how enzymes work.
3 What is the energy called that is needed to start a chemical reaction?
4 What do enzymes do to the level of activation energy required?
5 A lack of enzyme molecules will prevent the reaction from occurring any faster. What is this called?
6 Name four other factors that can affect the rate of an enzyme-catalysed reaction.
7 What is the difference between competitive and non-competitive inhibition?

inhibition: shape of the enzyme is changed, rate of reaction never reaches its maximum
inhibition; competitive inhibition: active site used by different molecule; non-competitive
4. lower it 5. limiting factor 6. concentration substrate, concentration enzyme; temperature, pH, competitive
1. lock and key 2. one enzyme for one job; reactions can go both ways, enzymes can be used over and over again 3. activation energy

Exam-style questions   Use the questions to test your progress. Check your answers on pages 94–95.

Biological molecules

1 a The equation shows two glucose molecules. Complete the equation to show what happens when two glucose molecules react together. [3]

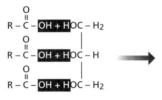

b Give the name of this type of reaction. [1]

...................................Hydrolysis...

c Name the bond that holds the two glucose molecules together. [2]

....glycosidic   bond·....................................

d Cellulose is made from long chains of β glucose molecules. Draw a diagram on separate paper to show the basic structure of a cellulose molecule. [1]

2 a Name three elements found in lipids. [3]

....C, H, O, oxygen...................................

b Describe how lipids differ structurally from carbohydrates. [1]

....

c Complete the following equation to show how a fatty acid and glycerol combine to form a lipid. [3]

$$
\begin{array}{l}
\text{O} \\
\| \\
\text{R} - \text{C} - \boxed{\text{OH} + \text{H}}\text{OC} - \text{H}_2 \\
\text{O} \\
\| \\
\text{R} - \text{C} - \boxed{\text{OH} + \text{H}}\text{OC} - \text{H} \\
\text{O} \\
\| \\
\text{R} - \text{C} - \boxed{\text{OH} + \text{H}}\text{OC} - \text{H}_2
\end{array}
\longrightarrow
$$

3 fatty acids   glycerol

d Name this type of reaction. [1]

...........................Hydrolysis...

e Name the bond that holds the fatty acid and glycerol together. [1]

.............................Glycosidic..

3 Describe the difference between saturated and unsaturated fats. [2]

....

4 a Phospholipids are used in cell membranes. Describe the steps in the process that cells use to make phospholipids. [3]

....
....
....

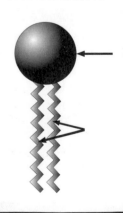

b The diagram shows a simplified phospholipid molecule. Label the diagram to describe the properties of the molecule. [4]

c Draw part of a cell membrane to show the positions of the phospholipid molecules. [2]

5 a List four elements that are sometimes found in protein molecules but not in carbohydrate or lipid molecules. [4]

....

b State which of these elements is always found in protein molecules. [1]

....

c Name the structural units that proteins are made from. [1]

....

6 a The diagram shows the structure of an amine acid. Label the amino group and the carboxylic acid group. [2]

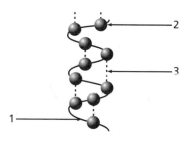

b Complete the equation to show how two amino acid molecules bond together. [2]

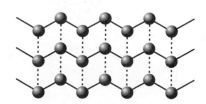

c Name this type of reaction. [1]

....

d Name the bond that holds the amino acids together. [1]

....

7 The diagram shows the secondary structure in a protein molecule.

a Label the arrows 1, 2 and 3. [3]

....

b Name this kind of secondary structure. [1]

....

8 The diagram shows another kind of secondary protein structure.

a State the advantage of this type of secondary structure. [1]

....

b On a separate sheet, draw a simple labelled diagram to show the structure of a fibrous protein. [2]

Total: /46

The ultra-structure of cells

The basic building blocks

● There are two main kinds of cell – animal cells and plant cells.

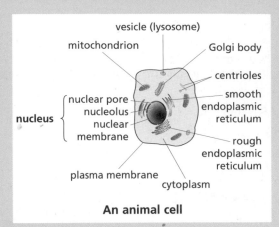

An animal cell

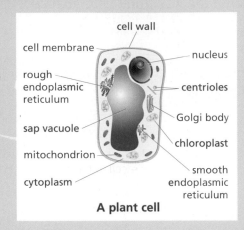

A plant cell

● Cells are the basic unit or building block for all living organisms.

● Animals and plants can have anything from one cell to millions of cells.

● Organisms that are made up of millions of cells have cells that are specialised to do all sorts of different jobs.

● Cells are more complicated than you might think. If we look at the cell's ultra-structure, we can see that they contain lots of different cell <u>organelles</u> that all have different functions.

Cell organelles

The cell membrane or plasma membrane

● The structure of the <u>cell membrane</u> is often explained using the <u>fluid-mosaic model</u>. This is because it is fluid and the protein molecules make it look like a mosaic.

● The cell membrane encloses the whole of the cell.

● It protects the cell from the outside environment.

● It allows certain substances to enter and leave the cell through protein channels.

● The membrane has **two layers** of lipid (fat) molecules.

Look at the way the lipid molecules are arranged. The hydrophilic head ends 'love' water. They face the outside of the membrane. The hydrophobic tail ends 'hate' water. They face the inside of the membrane.

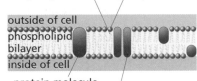

phospholipid molecule =
hydrophobic tail / hydrophilic head

outside of cell
phospholipid bilayer
inside of cell

protein molecule
protein channel to transport specific substances into cell

The structure of a cell membrane

nucleolus (RNA and ribosomes made here)

nuclear pore (mRNA moves out here)

Cell nucleus

Cytoplasm

● Each cell is filled with **cytoplasm**.

● Many substances are dissolved in it and many chemical reactions take place there.

Nucleus

● The nucleus contains the **DNA** that stores all the genetic information.

● The DNA codes for all the proteins and is kept safe from all the chemical reactions taking place outside the nucleus in the cytoplasm.

● The nucleus has pores through which RNA carries the instructions from the DNA out into the cell to make proteins.

Cell organelles

Mitochondria

The **mitochondria** are sometimes called the powerhouse of the cell. It is where energy is released from glucose.

A mitochondrion

- The inner fluid matrix has many folded membranes called **cristae**.
- These provide a large surface area for carrying out the biochemical processes of respiration.
- This releases energy for the cell in the form of ATP.

Ribosomes

- **Ribosomes** are where the proteins are made.
- RNA from the nucleus provides codes for making proteins.

Golgi body

- The **Golgi body** is involved with the production of substances such as carbohydrates and glycoproteins.
- The substances are formed inside the membrane of the Golgi body and then become isolated in buds or **vesicles**.
- The vesicles can merge with the cell membrane so the cell can secrete these substances.

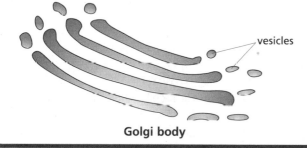

Golgi body

Centrioles

- **Centrioles** are bundles of **microtubules**.
- They are used when the cell divides.
- During cell division the centrioles move to opposite ends of the cell as the spindle is formed.

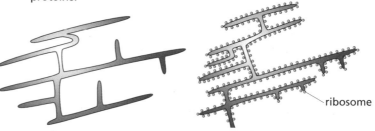

Centrioles

Endoplasmic reticulum

- The **endoplasmic reticulum** (ER) has channels in which materials are transported around the cell.
- Some ER is smooth. It helps in the manufacture and transport of lipids.
- Some ER is rough. It has ribosomes which manufacture proteins.

Smooth endoplasmic reticulum   **Rough endoplasmic reticulum**

Lysosomes

- **Lysosomes** contain digestive enzymes such as lysozyme.
- They are responsible for breaking down unwanted parts of the cell.
- It is important that the enzymes stay inside the membranes of the lysosome. If they escaped they would destroy the cell.

Lysosome

HINT Chefs sometimes hammer a beef steak because it breaks open the lysosomes and releases the enzymes. The enzymes start to break down the meat and tenderise it.

Quick test

1 **Why is the cell membrane sometimes explained using a fluid-mosaic model?**
2 **What does the cell membrane do?**
3 **Describe the function of the nucleus.**
4 **Describe the function of the mitochondria.**
5 **Describe the function of the ribosomes.**
6 **Describe the function of the endoplasmic reticulum.**
7 **Describe the function of the Golgi body.**
8 **Describe the function of the centrioles.**
9 **Describe the function of the lysosomes.**

1. It is fluid and it looks like a mosaic. 2. It contains what enters and organises the cell. 3.It contains and protects the DNA. 4. They provide the cell with energy. 5. They produce the cell's proteins. 6. It transports materials around the cell. 7. It secretes substances such as glycoproteins in its vesicles. 8. They help to produce the spindle when the cell divides. 9. They contain digestive enzymes to break down unwanted substances.

Plant cells

Some structures are found only in plant cells, and not in animal cells. The three extra structures are:

● a cell wall
● a vacuole
● chloroplasts.

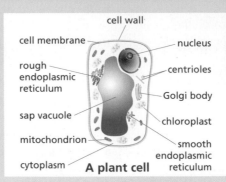

A plant cell

(labels: cell wall, cell membrane, nucleus, rough endoplasmic reticulum, centrioles, Golgi body, sap vacuole, chloroplast, mitochondrion, smooth endoplasmic reticulum, cytoplasm)

EXAMINER'S TOP TIP
Make sure you know the differences between an animal cell and a plant cell and can explain what each of the extra features does.

Cell wall

● The cell wall is made of **cellulose**.
● It surrounds the plasma membrane.
● It gives the cell extra support by maintaining the structure of the cell against the internal hydrostatic pressure.

HINT Think of a bicycle tyre. The inner tube is the plasma membrane. The outer tyre is the cell wall. When the inner tube is inflated the outer tyre is hard and firm.

● The cytoplasm from one cell is connected to the cytoplasm of the neighbouring cell by connections through the cell wall called **plasmodesmata**.

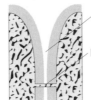

(labels: middle lamella, plasmodesmata (a strand of cytoplasm connected to next cell))

Plasmodesmata

● The cell walls of neighbouring cells are joined by the **middle lamella**.

Cell vacuole

● The vacuole is filled with sap.
● The sap contains chemicals such as mineral ions and glucose dissolved in water.
● The membrane around the vacuole is called the **tonoplast**.
● Water enters the vacuole by osmosis.
● This helps the cell to produce an internal hydrostatic pressure.
● This pressure keeps the cell turgid and helps provide support for the plant.

Chloroplasts

● The **chloroplasts** are where photosynthesis takes place to convert water and carbon dioxide into oxygen and glucose.

(labels: outer membranes, stroma, granum (a stack of membranes containing chlorophyll), thylakoid membranes)

A chloroplast

● The chloroplast is bounded by two membranes.
● Inside the chloroplast, other membranes, called **thylakoid membranes**, form stacks called **grana** (singular – **granum**). These membranes contain the chlorophyll.
● The material between the grana is called **stroma**.

EXAMINER'S TOP TIP
Make sure you know the differences between prokaryotes and eukaryotes.

Prokaryotes and eukaryotes

Eukaryotes are organisms that have a cell structure containing all the cell organelles described on the last two pages. Most importantly, eukaryotes have a nucleus.
Prokaryotes are more primitive organisms that do not possess all of the cell organelles. They have a much simpler structure. Examples of prokaryotes are **bacteria** and **blue–green algae**. Scientists think that prokaryotes evolved before the more complex eukaryotes.

Prokaryotes:
● do not have their DNA enclosed in a nucleus
● have a cell wall but it consists of mucopeptides, not cellulose
● do not have mitochondria, Golgi bodies or endoplasmic reticulum.
● have small ribosomes.

HINT When using a microscope to look at organisms, check to see whether the cells have nuclei. If they do not, they are prokaryotes.

The bacterium does not have a nucleus. The DNA is free inside the cytoplasm.

Separation of cell organelles

Separation using cell fractionation

When scientists want to study the organelles found inside the cells, they use a process called <u>cell fractionation</u>.

Stage 1 – Preparing the cells

- The cells are placed in an <u>**isotonic solution**</u>. This is a solution with a concentration equal to the concentration of the cell. This ensures that there is no net flow of water into or out of the cell.
- The solution contains a buffer. This ensures that the pH remains constant. Any change to the pH could damage the cell organelles.

- The cells and solution are then cooled down to about 5 °C.
- This slows down all the metabolic activity in the cells. Think about what would happen when the lysosomes were damaged if the cells were not cooled down.

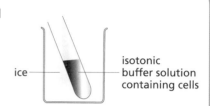

ice — isotonic buffer solution containing cells

Stage 2 – Homogenising the cells

- Cell walls and membranes need to be broken open to release the organelles from inside the cell.
- The solution containing the cells is placed inside a blender.

- Although some organelles will be damaged by the blender, many will survive unscathed.
- The liquid now consists of a suspension containing the organelles.

blender

Stage 3 – Separating the organelles

- The suspension is placed in a centrifuge.
- The centrifuge spins so fast it is often called an <u>**ultracentrifuge**</u>.
- Because the different organelles have a different mass, they will move to the bottom of the centrifuge tubes at different speeds.
- The heaviest organelles, such as nuclei, will move towards the bottom of the tube first.

First spin   **After spin**

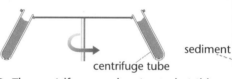

centrifuge tube

supernatant { with cytoplasm, ER, mitochondria, ribosomes

sediment --- nuclei

- The centrifuge can be stopped at this point and the supernatant liquid removed using a pipette, leaving the nuclei behind.
- The suspension can then be centrifuged again to remove the lighter mitochondria.

- Finally it can be spun once more to remove the even lighter ribosomes.

Quick test

1 List three differences between plant cells and animal cells.
2 Describe the significance of each of these differences.
3 Suggest why mitochondria and chloroplasts have large internal membranes.
4 Explain the difference between prokaryotic and eukaryotic organisms.
5 Suggest why scientists think that prokaryotic organisms evolved before eukaryotic ones.
6 Name two groups of prokaryotic organisms.
7 Name the three stages in separating organelles from their cells.
8 Explain why cells are cooled down before the organelles are separated.
9 Explain why nuclei separate out before mitochondria during ultracentrifuge.

1. Plant cells have chloroplasts, a cell wall and a vacuole. 2. Chloroplasts enable the plant to photosynthesise and make food; the cell wall provides support; the vacuole can provide storage turgor to the cell. 3. to provide a surface for enzymes and metabolic reactions. 4. Prokaryotic organisms do not have a nucleus, an endoplasmic reticulum, Golgi bodies, or mitochondria. 5. They are less specialised, therefore evolved first. 6. bacteria and blue–green algae. 7. preparing the cells; homogenising the cells; separating the organelles 8. to stop metabolic activity, including that of the enzymes in the lysosomes which would destroy the organelles; increase lifespan of organelle 9. They have a greater mass.

Microscopes

There are two main kinds of microscope.

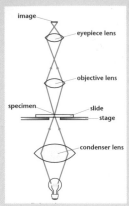

Light microscope

A <u>light microscope</u>:

- uses light
- has an eyepiece through which material is viewed
- can be used to view living material
- can magnify up to a thousand times depending on the lenses that are used.

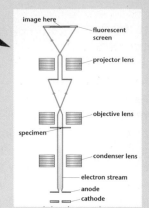

Transmission electron microscope

An <u>electron microscope</u>:

- uses electrons
- is viewed on a screen
- cannot be used to view living material
- can magnify many thousands of times.

Electronmicrographs of cell organelles

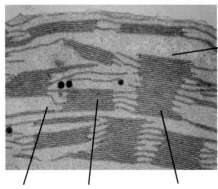

mitochondria

Golgi body   nuclear membrane   endoplasmic reticulum

ED - No indication of which image is which, and placing of labels, etc

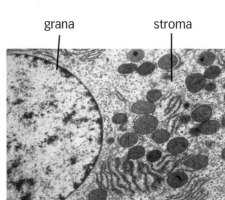

grana   stroma

mitochondrion

cristae

EXAMINER'S TOP TIP

Learn to identify all the cell organelles in the electronmicrographs.

Things to do with a light microscope

1. Measure size

A <u>**graticule**</u> helps you to calculate the actual size of a specimen.

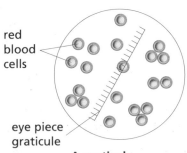

red blood cells

eye piece graticule

A graticule

EXAMINER'S TOP TIP

You should know how to carry out both of these tasks.

2. Count numbers

A <u>**haemocytometer**</u> allows you to count the number of red blood cells in a given volume of blood.

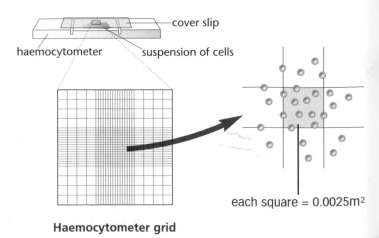

cover slip

haemocytometer   suspension of cells

each square = 0.0025m²

Haemocytometer grid

Cell specialisation

Many organisms consist of millions of cells. In most of these multicellular organisms, the cells are <u>specialised</u>. Different groups of cells do different jobs. Specialisation of cells makes organisms more efficient and enables them to become adapted to living in different environments.

HINT Think of a person living on a desert island compared with someone living in a modern society.

- On a desert island you do everything for yourself: catch food, build a shelter, get firewood.
- If other people lived with you, each one could do a different job for everyone else.
- If you lived in a city, you would specialise to learn one job. For example, you might be a chef. You would prepare food for lots of other people. Other people would specialise to do other jobs. They would provide you with electricity and clean water. They will even build a car for you. The list is endless. You would have a much better quality of life.

Red blood cells

- Red blood cells have no nucleus.
- This means they can have a large surface area compared to their volume.

A red blood cell

- They are red because they contain **haemoglobin**. Haemoglobin combines with oxygen in the lungs and releases the oxygen to the tissues that need it.

Guard cells

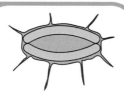

- Guard cells open and close the stomata on the undersurface of a leaf.

A guard cell

- As their internal turgor increases, their shape changes and so the stomata open.

Nettle stinging cell

- Plants like stinging nettles have cells specialised for stinging.

Motor neurones

- Motor neurones are long and thin.
- They carry electrical impulses from the brain to the muscles of the body.

A motor neurone

Tissues and organs

When cells are specialised, they also need to be organised. This means that cells that do the same job are grouped together. However, there is more than one level of organisation.

Tissues
A <u>tissue</u> is a collection of similar cells that work together to carry out a function. Muscle cells work together to contract and make parts of the body move.

Organs
An <u>organ</u> is a collection of tissues which work together. The stomach consists of muscle and other kinds of tissue. They work together to digest food.

Organ systems
A variety of organs and tissues combines to make an <u>organ system</u>. The digestive system consists of the stomach and gut plus various glands.

EXAMINER'S TOP TIP
A common type of exam question involves asking why specialisation leads to greater efficiency.

Quick test

1. List four differences between a light microscope and an electron microscope.
2. List two other things that a light microscope can be used for other than viewing specimens.
3. What are the advantages for organisms having specialised cells?
4. How is a red blood cell specialised to do its job?
5. Put the following words in their correct order, starting with the simplest: tissue, organ system, cell, organ.

1. An electron microscope uses electrons not light; the specimen is viewed on a screen; it can only be used for dead material; the magnification is greater. 2. measuring sizes and counting numbers 3. It allows cells to be used more efficiently and the organism to be adapted to its environment. 4. It has large surface area compared to volume and contains haemoglobin to carry oxygen. 5. cell, tissue, organ, organ system

Transport in cells

Why are organisms multicellular?

- Some organisms are composed of a single cell.
- All the substances that the cell needs for life must pass in through the cell plasma membrane.
- All the waste materials that the cell produces must pass out through the plasma membrane.
- This creates a potential problem: as cells increase in size, the volume of the cell increases more than the surface area.
- In a sphere, the volume increases to the power cubed; the surface only increases to the power squared.

- This limits the size of a cell. The bigger the cell gets, the less the surface area (plasma membrane) is able to supply the cell with all its needs. The cell will die.
- If a single-celled organism is to become larger, the only solution is to be composed of more than one cell. This is why organisms became <u>multicellular</u>.

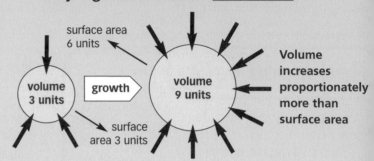

surface area 6 units

volume 3 units

growth

volume 9 units

surface area 3 units

Volume increases proportionately more than surface area

Plasma membranes

The entry and exit of substances into and out of cells is controlled by the **plasma membrane**.

protein channel to transport specific substances into cell

outside of cell

extrinsic protein molecule

phospholipid bilayer

hydrophobic tail

inside of cell   intrinsic protein molecule

phospholipid molecule

hydrophilic head

Plasma membrane

- The membrane is a <u>**bilayer**</u> consisting of <u>**phospholipid molecules**</u>. The hydrophilic heads point outwards. They attract water molecules.
- The hydrophobic tails point inwards. They repel water molecules.
- Embedded in the bilayer are protein molecules. <u>**Extrinsic protein molecules**</u> are only embedded in the surface of the membrane. <u>**Intrinsic protein molecules**</u> pass all the way through the membrane.

The different molecules of the cell membrane have different functions in the transfer of substances across the membrane.

EXAMINER'S TOP TIP

Make sure you can label a diagram of a plasma membrane and know what each type of molecule does.

Carrier proteins

<u>**Carrier proteins**</u> use energy from ATP to carry molecules across the membrane.

Carrier proteins have a specific shape. They carry molecules with a complementary shape across the membrane.

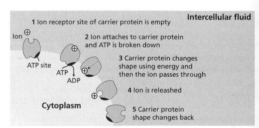

1 Ion receptor site of carrier protein is empty   **Intercellular fluid**

Ion

2 Ion attaches to carrier protein and ATP is broken down

ATP site

ATP
ADP

3 Carrier protein changes shape using energy and then the ion passes through

4 Ion is released

Cytoplasm

5 Carrier protein shape changes back

A carrier protein molecule taking a molecule across a cell membrane

Phospholipid molecules

small lipid-soluble molecules pass through

Lipid molecules and small molecules with no charge pass through the membrane.

Channel proteins type 1 – pores

Charged ions and larger molecules such as glucose cannot pass through the phospholipid layer. Instead they pass through permanently opened **pore channel proteins.**
- Channel proteins are specific.
- Each channel protein will only let one type of molecule through.
- Water can also enter the cell through channel proteins.

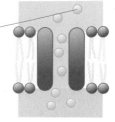

glucose molecules

Pore channel protein

Channel proteins type 2 – gated

Gated channel proteins can be either open or closed.
- The gated channel protein has a receptor site.
- A hormone binds with the receptor site to open the channel.
- A specific molecule can then pass through and enter the cell.

When there is a lot of glucose in the blood after a meal, the hormone insulin opens gated channels in cells to allow them to absorb more glucose. Thus the glucose level in the blood is lowered.

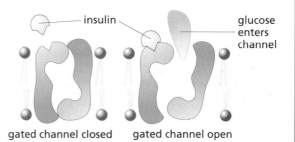

insulin

glucose enters channel

gated channel closed   gated channel open

Gated channel protein

Recognition proteins or antigens

Some extrinsic proteins have a carbohydrate on their outer surface. These carbohydrate–protein molecules, or **glycoproteins**, help with cell recognition. They are called **antigens**. The cells in a body recognise antigens belonging to other cells in that body. But any foreign organism that invades will carry foreign antigens. The body's cells will not recognise them and will attack them with defence cells.

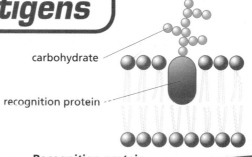

carbohydrate

recognition protein

Recognition protein

Quick test

1 *What is the relationship between surface area and volume?*
2 *How are large organisms possible if the size of a cell is limited?*
3 *Name the protein molecule that is embedded in the surface of the plasma membrane.*
4 *Name the protein molecule that spans the width of the plasma membrane.*
5 *What kinds of molecules can pass through the phospholipid layer?*
6 *Which protein molecules use the energy from ATP to transport other molecules across the plasma membrane?*
7 *Name two kinds of channel protein.*
8 *Which channel protein is permanently open?*
9 *Which channel protein is opened by a hormone?*
10 *Which type of extrinsic protein can be a glycoprotein?*
11 *What are glycoproteins made from?*

1. Volume increases to the power cubed; surface area increases to power squared. 2. They have to be multicellular. 3. extrinsic 4. intrinsic 5. lipids and small uncharged molecules. 6. carrier proteins 7. pore and gated 8. pore 9 gated 10. recognition protein or antigen. 11. protein and carbohydrate

Movement into and out of cells

- <u>Diffusion</u> is the movement of molecules from where there are many to where there are few.
- We say the molecules are moving <u>along (or down) a concentration gradient</u>.
- Diffusion is a <u>passive process</u>. This means it does not require energy to be used.
- It can happen in living and non-living systems.
- Diffusion continues until the molecules are evenly distributed. Then we say they are in <u>equilibrium</u>.

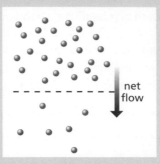

net flow

Movement of molecules in diffusion

Diffusion works faster when …

- the temperature is higher – molecules move faster because they have higher kinetic energy
- there is a larger surface area – for molecules to diffuse through
- the molecules are smaller
- there is a greater diffusion gradient.

Facilitated diffusion

Some molecules cannot easily diffuse through a plasma membrane. Channel proteins can help (facilitate) move them through the membrane. This is called **facilitated diffusion**.

- A molecule binds with a channel protein.
- The channel protein changes shape.
- The molecule is released into the inside of the cell.
- Although no energy is used in the process, it is much faster than normal diffusion.

once in position the molecule changes the shape of the carrier protein

the site gives up the molecule on the inside of the cell

carrier protein

Channel protein helps a molecule through the membrane

EXAMINER'S TOP TIP
Make sure you know the difference between diffusion, facilitated diffusion and osmosis.

Active transport

- <u>Active transport</u> requires the use of energy from ATP.
- Molecules are moved **against (or up) a concentration gradient**. This means that molecules are moved from an area of lower concentration to an area of higher concentration.
- In neurones, carrier protein molecules transport sodium ions out of the cell and potassium ions into the cell.
- The carrier proteins are sometimes called **protein pumps**.
- Some plant nutrients are absorbed by active transport.

Pinocytosis

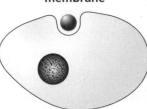

substance touches cell membrane

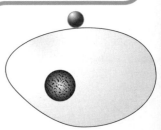

membrane surrounds the substance

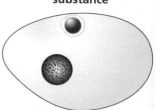

substance enters the cell in a vacuole

Osmosis

<u>Osmosis</u> is a special kind of diffusion. Water molecules diffuse across a **selectively permeable membrane**.
Only water molecules can move through the membrane.

- Water molecules move from where there are lots of water molecules to where there are few.
- Water molecules move from where there are few solute molecules to where there are many.

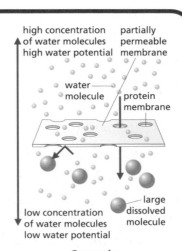

high concentration of water molecules high water potential

partially permeable membrane

water molecule

protein membrane

low concentration of water molecules low water potential

large dissolved molecule

Osmosis

Osmosis and water potential

- The greater the difference in the number of water molecules on either side of the membrane, the greater the water potential.
- Water potential is represented by the Greek letter ψ (psi, pronounced 'sigh').

Pure water has a water potential of 0 kPa. As substances dissolve in the water, the water potential drops and becomes negative. Water moves by osmosis to the region with the lowest water potential.

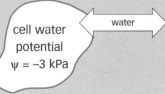

Surrounding water = −3 kPa potential

cell water potential ψ = −4 kPa — enters cell water

cell water potential ψ = −3 kPa ← water →

cell water potential ψ = −2 kPa — leaves cell water

1 Cell in a **hypotonic** solution – one that is less concentrated (higher water potential).

2 Cell in an **isotonic** solution – solutions on either side of the membrane have equal concentration.

3 Cell in a **hypertonic** solution – one that is more concentrated (lower water potential).

Water potential equation

water potential of cell	=	solute potential of ions inside cell	+	pressure potential of cell wall
ψ cell		ψ s		ψ p
almost always negative		always negative		usually positive

Osmosis in animal cells

- If an animal cell is placed in distilled water, it swells and bursts.
- If an animal cell is placed in hypertonic solution, it shrinks as it loses water.

 Normal blood cell

 Cell in hypertonic solution

Cell in hypotonic solution

Osmosis in plant cells

- If a plant cell is placed in distilled water it becomes **turgid**. The cell wall stops it from bursting.
- If a plant cell is placed in a hypertonic solution it **plasmolyses** as the cell membrane pulls away from the cell wall.

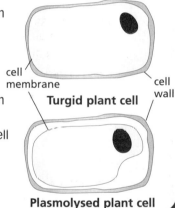

cell membrane

cell wall

Turgid plant cell

Plasmolysed plant cell

Quick test

1 *Under what conditions does diffusion work faster?*
2 *Explain the differences between diffusion and facilitated diffusion.*
3 *Explain how active transport is different from facilitated diffusion.*
4 *Describe the process of pinocytosis.*
5 *Explain the differences between osmosis and diffusion.*
6 *In what units is water potential measured?*
7 *What is the water potential of pure water?*
8 *What happens to a blood cell and a plant cell when placed in a hypertonic solution?*

1. higher temperature, larger surface area, smaller molecules, larger diffusion gradient. 2. Facilitated diffusion uses channel proteins to speed up the movement of molecules into the cell; it is faster than normal diffusion. 3. Active transport uses energy from ATP to carry molecules into the cell against a concentration gradient. 4. The membrane surrounds the molecule to be taken into the cell. The membrane then detaches from the cell plasma membrane to form a vacuole containing the molecule inside the cell. 5. Osmosis is the movement of water molecules across a selectively permeable membrane from an area of high water potential to an area of low water potential. 6. kPa 7. 0 kPa 8. A blood cell bursts; a plant cell becomes turgid.

Cells

1 Explain why the cell membrane is sometimes referred to as the fluid-mosaic model. [2]

....

2 The diagrams show cell organelles.
   a Identify each of the structures. [4]

| 1 | 2 | 3 | 4 |

....

   b Describe the role of each of the structures. [4]

....
....

3 Describe how the structure of prokaryotic organisms differs from eukaryotic organisms. [4]

....
....

4 Plant cell walls are more than just a cellulose box.
   Describe how the cell wall is adapted to the jobs that it does. [2]

....
....

5 The diagram shows the internal structure of a chloroplast.
   a Label the arrows on the diagram. [3]

1. ....

2. ....

3. ....

   b Name the part of the chloroplast where light energy is absorbed for photosynthesis. [1]

....

6 Describe the process for extracting cell organelles from cells and separating them. [3]

....
....

7 a Describe the differences between a light microscope and an electron microscope. [4]

....

b Label the arrows on the diagram of a light microscope. [4]

1. ....

2. ....

3. ....

4. ....

8 Explain why large organisms are composed of many cells. [2]

....
....

9 Label the two protein molecules in the diagram. [2]

....

10 a Use the diagram to explain how channel proteins (pores) allow glucose molecules to cross the plasma membrane. [2]

....
....

b Use the diagram to explain how gated channel proteins allow glucose molecules to cross the plasma membrane. [3]

....
....
....
....

11 The diagram shows an extrinsic protein molecule in the plasma membrane. Explain what is different about this protein molecule. [3]

....
....

12 Explain the differences between:
a diffusion ....
b facilitated diffusion ....
c active transport ....
d pinocytosis .... [4]

13 a Write down the equation for water potential.

....

b Copy out these cells and draw arrows to show the direction of net water movement between the cells.

−400 kPa

−600 kPa   −500 kPa   [3]

Total: /51

27

Gaseous exchange

Systems for gas exchange

Larger, more complex organisms have a larger volume to surface area ratio. This means they need special organ systems to deal with gaseous exchange so that all the cells of the organism are provided for.

Gas exchange in a leaf

Plants exchange gases through their leaves. They take in carbon dioxide for photosynthesis and release oxygen. The leaf structure is specially adapted to do this.

- Leaves are very thin to allow gases to diffuse through.
- There are large air spaces between cells to allow gases to circulate.
- Cells have a large surface area for gases to diffuse through.
- Holes on the leaf surface called **stomata** allow gases to enter and leave the leaf.
- Specialised cells called **guard cells** can open and close the stomata.

Section through a leaf

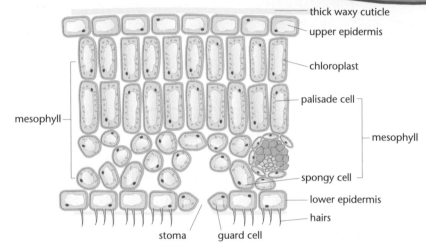

- thick waxy cuticle
- upper epidermis
- chloroplast
- palisade cell
- mesophyll
- mesophyll
- spongy cell
- lower epidermis
- hairs
- stoma
- guard cell

Gas exchange in a fish

- Gas exchange in a fish takes place through the fish's gills.
- Each gill has many **gill lamellae**.
- Each lamella is covered with **gill plates**.
- Gill plates are very thin, have a large surface area and have lots of blood vessels.
- The blood flows in the opposite direction to the water.
- This is called **counter current**.
- It maintains a higher concentration gradient to exchange gases.

Gill bar (bone)
Gill lamellae
Capillaries
Gill bar (bone)
Gill lamellae
A gill
blood vessels

Con current

This is less efficient as only 50% of gases can be absorbed.

| → | 90 | 80 | 70 | 60 | 50 | 50 | 50 | 50 | 50 | water |
| → | 10 | 20 | 30 | 40 | 50 | 50 | 50 | 50 | 50 | blood |

Counter current

This is very efficient – a higher percentage of the gases can be absorbed.

| → | 100 | 90 | 80 | 70 | 60 | 50 | 40 | 30 | 20 | 10 | water |
| ← | 90 | 80 | 70 | 60 | 50 | 40 | 30 | 20 | 10 | 0 | blood |

Respiratory surfaces

Remember, respiratory surfaces need to be:

Thin

moist – to speed up diffusion

and they need to have:

a large surface area

a good transport supply

Gas exchange in mammals

In mammals, air has to be taken into the lungs for gas exchange to take place.

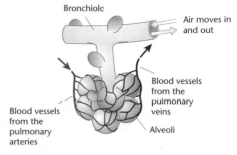

Inhaling (breathing in)

1. The internal intercostal muscles relax; external intercostal muscles extract.
2. The diaphragm contracts and flattens.
3. Both of these actions will increase the volume in the pleural cavity and so decrease the pressure.
4. Air is therefore drawn into the lungs.

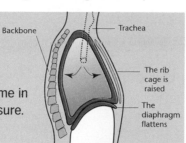

Exhaling (breathing out)

1. The internal intercostals contract; external intercostals relax.
2. The diaphragm relaxes and domes upwards.
3. The volume in the pleural cavity is decreased so the pressure is increased.
4. Air is forced out of the lungs.

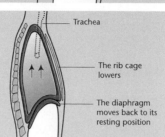

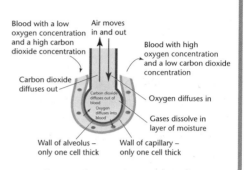

Blood supply to the alveoli

Gas exchange in an alveolus

Transport systems in mammals

- Mammals have a **double circulatory system**.
- This means that the blood goes through the heart twice on each circuit of the body.
- This double pump enables the blood to carry substances around the body more efficiently.
- The system is closed. This means that all the blood travels around the body inside blood vessels.
- Mammals are warm blooded and use large amounts of energy to survive. A large pump like the heart and an efficient transport system are vital to supply the cells of the body with all that they need.

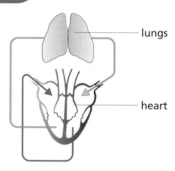

Double circulation system enclosed in blood vessels

Quick test

1. *Why do organisms need complex methods of gaseous exchange as they get bigger?*
2. *How is a leaf adapted for gaseous exchange?*
3. *Why does water flow in the opposite direction to the blood in fishes' gills?*
4. *Why do gill lamellae have such a good blood supply?*
5. *What do all respiratory surfaces have in common?*
6. *Describe the path taken by an oxygen molecule as it enters the lungs.*
7. *Describe how air is drawn into the lungs.*
8. *Why is the heart divided into two pumps?*
9. *What is meant by a closed circulatory system?*

1. The volume increases faster than the surface area and all the cells need to be supplied. 2. thin structure, lots of air spaces between cells, stomata bordered by guard cells, large internal surface area for cells 3. Counter current flow ensures greater absorption of oxygen from the water and release of carbon dioxide back to the water. 4. to absorb all the oxygen and transport it around the fish's body OR to carry all the carbon dioxide from the cells of the fish back to the gills 5. large surface area, moist, thin, good transport system 6. nose, trachea, bronchi, bronchioles, alveoli 7. diaphragm contracts and lowers, ribs are raised, internal volume of pleural cavity increases, pressure drops, air enters 8. to provide extra pumping power to get the blood around the body more quickly 9. The blood is kept inside blood vessels.

The heart

The mammalian heart consists of four chambers.

Blood flows from the top chambers to the bottom ones.

The blood cannot get directly from one side to the other without leaving the heart and returning to it.

- The <u>bicuspid</u> and <u>tricuspid valves</u> (<u>atrioventricular valves</u>) link the <u>atria</u> to the <u>ventricles</u> and stop blood flowing back.

- <u>Chordae tendinae</u> hold the valves in place.

- <u>Semi-lunar valves</u> stop blood flowing back into the heart.

- The heart is made from <u>cardiac muscle</u>. Cardiac muscle cells can contract and relax without any external nerve stimulus. This type of muscle is called <u>myogenic</u>.

- The heartbeat is controlled by a part of the brain called the <u>medulla oblongata</u>.

- The <u>sympathetic nerve</u> increases the rate of heartbeat.

- The <u>parasympathetic nerve</u> (<u>vagus</u>) decreases the rate of heartbeat.

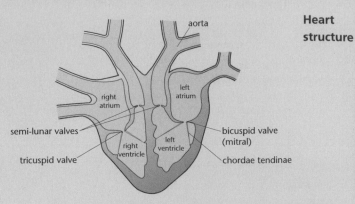

Heart structure

aorta

right atrium

left atrium

semi-lunar valves

bicuspid valve (mitral)

tricuspid valve

right ventricle

left ventricle

chordae tendinae

Control of the heartbeat

nerves from the medulla go to the **sino-atrial node** causing right and then left atrium to contract

right atrium

left atrium

atrio-ventricular node

bundle of His contains **Purkinje fibres** – these carry electrical activity to ventricles causing them to contract

right ventricle

left ventricle

atrio-ventricular node then conducts electrical activity through **bundle of His**

EXAMINER'S TOP TIP
The structure of the heart, cardiac cycle and control of heartbeat should be carefully learned. It is a favourite with examiners.

The cardiac cycle

1

atria contract – systole

valves pushed open

ventricles relax – diastole

atria relax – diastole

ventricles contract – systole

semi-lunar valves oper

valves pushed closed

2

3

atria and ventricles relax – diastole

higher pressure in blood vessels results in blood filling atria

Look at these diagrams carefully.
Remember – **systole** = contract
diastole = relax

- The whole of the **cardiac cycle** is completed in less than a second.
- Too much carbon dioxide in the blood speeds up the heartbeat.
- Too little carbon dioxide in the blood slows the heartbeat down.
- The level of carbon dioxide in the blood is monitored by sensors in the wall of the aorta and carotid artery.
- An abnormal rhythm in the heartbeat can result in death owing to an inability to pump blood.

Blood pressure during the cardiac cycle

The graph shows changes in pressure in various parts of the heart during the cardiac cycle.

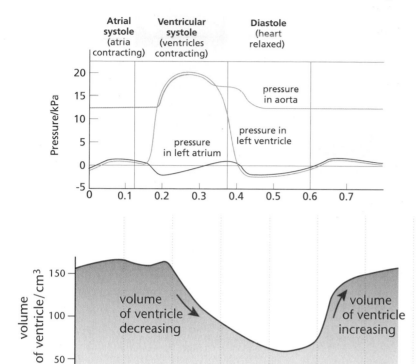

EXAMINER'S TOP TIP
Examiners often ask questions that require you to describe what is happening at various points on the graph

EXAMINER'S TOP TIP
Examiners also like asking you to relate diagrams to various parts of the graph. Note – pressure is measured in kPa.

- Look at how the volume of blood in the ventricles relates to the pressure in the ventricles.
- Notice that when the pressure increases, the volume decreases.
- Notice that at 0.4 seconds, when the pressure in the atria is at its peak, the volume in the ventricles starts to rise.

Breathing rate and heartbeat rate are closely linked. When one increases, so does the other. The sino-atrial node, the atrio-ventricular node, the bundle of His and Purkinje fibres form the heart's **pacemaker**.
Some people need an artificial pacemaker to keep their heart beating.

Pacemaker

Quick test

1 **What does the word myogenic mean?**
2 **What part of the brain controls heartbeat rate?**
3 **What type of nerve increases heartbeat?**
4 **What are valves between the atria and the ventricles called?**
5 **Name the parts of the heart's pacemaker mechanism.**
6 **What does systole mean?**
7 **What does diastole mean?**
8 **What type of valve prevents the backflow of blood from the aorta?**
9 **How long is the cardiac cycle? Look at the graph at the top of the page.**
10 **What happens to the volume of blood in a ventricle when the atrium contracts?**

1. the natural ability of the cardiac muscle cell to contract rhythmically 2. medulla oblongata 3. sympathetic nerve 4. atrio-ventricular valves 5. sino-atrial node, atrio-ventricular node, Purkinje fibres, bundle of His. 6. contraction of heart muscle 7. relaxation of heart muscle 8. semi-lunar 9. less than 1 second 10. It increases.

Blood vessels

- Blood is transported around the body in blood vessels.
- Blood leaving the heart travels through <u>arteries</u>.
- Blood returning to the heart travels through <u>veins</u>.
- Blood is taken to the tissues in <u>capillaries</u>.

Arteries

- Blood leaves the heart through **arteries**.
- Blood is under high pressure in the arteries because they are close to the heart.
- During diastole, the smooth muscle in the **tunica media** contracts.
- This ensures that blood pressure is maintained between heartbeats and keeps the blood flowing.
- **Tunica externa** forms a stretchy outer layer made from **collagen fibres**. Flat **endothelial cells** form the smooth inner lining.

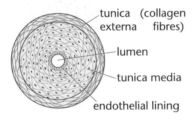

tunica (collagen externa fibres)
lumen
tunica media
endothelial lining

Veins

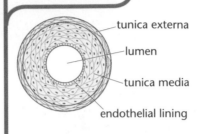

tunica externa
lumen
tunica media
endothelial lining

direction of blood flow

semi-lunar valve

- The **tunica media** is thinner in veins than it is in the arteries as the blood pressure is lower.
- Veins have **semi-lunar valves** to prevent the backflow of blood.
- Blood pressure is lowest in the veins as they are furthest from the heart.
- Veins cannot do anything to move the blood to the heart. When we move, contraction of skeletal muscle squeezes the veins and pushes the blood upwards towards the heart.

Blood pressure

- A normal person's blood pressure is about 120 over 80.
- The first figure is the pressure created during **systole**.
- The second figure is the pressure maintained by the arteries during **diastole**
- As we get older and the arteries harden, the first figure rises because the arteries cannot expand to absorb the pressure of systole.

Capillaries

- The blood flows from the arteries into capillaries.
- Capillaries are much smaller and thinner than arteries.
- Capillaries have a very narrow diameter, which slows down the blood flow.
- Capillary walls are only one cell thick, so substances are exchanged between the blood and tissue cells very easily.
- There are thousands of miles of capillaries in the human body.

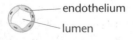

endothelium
lumen

What is in blood?

Red blood cells with haemoglobin for transporting oxygen.

cytoplasm with large amount of haemoglobin

shape gives a large surface area

front   side

Lymphocytes (white blood cells) make proteins called antibodies that kill invading cells.

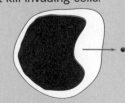

Phagocytes (white blood cells) engulf foreign cells such as bacteria.

Platelets are small cell fragments that help blood to clot.

Not to scale

Red blood cells

- People who live at high altitude have more red blood cells than people who live at sea level.
- This is because there is less oxygen at high altitude. So the bone marrow makes more red blood cells so that the blood can take up enough oxygen.
- Some athletes train at high altitude. Their bone marrow makes more red blood cells so that when they return to a lower altitude, their blood can carry more oxygen to their muscles. This helps them perform better.

Plasma, tissue fluid and lymph

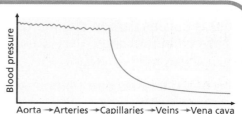

Hydrostatic pressure in blood vessels

- When blood enters capillaries, it is under high **hydrostatic pressure** (HP).
- Because the capillaries are so thin, they leak small molecules such as water and glucose. This is called **ultrafiltration**.
- Protein molecules are too large and stay behind in the blood.
- The liquid that collects in the intercellular spaces is called **tissue fluid**.
- This leaking of molecules causes the hydrostatic pressure in the capillaries to drop.
- Further along the capillary, the hydrostatic pressure becomes so low that some of the water molecules start to re-enter the capillaries by osmosis.
- This is because the **water potential** (WP) in the tissue fluid is now higher than the hydrostatic pressure in the blood.
- The tissue fluid that is left is collected by **lymph vessels**.
- Once in the lymph vessels, the tissue fluid is called **lymph**.
- The lymph vessels eventually return the lymph back into the blood.

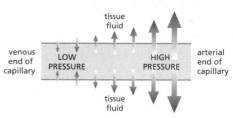

Ultrafiltration in a capillary

The job of tissue fluid
- The tissue fluid supplies the cells with oxygen and glucose.
- Oxygen and glucose diffuse into the cells along a concentration gradient (see active transport and facilitated diffusion on p. 24).
- Carbon dioxide diffuses out of the cells.

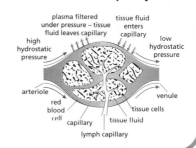

Quick test

1 Name the different tissues found in an artery and explain the purpose of each.
2 How is the structure of a capillary related to its function?
3 How is the structure of a vein related to its function?
4 Explain the role of phagocytes and lymphocytes.
5 Why do people who live at high altitude have more red blood cells per mm³ of blood?
6 What is the process called where high hydrostatic blood pressure forces small molecules out through the capillaries?
7 Why is tissue fluid only formed at the arterial end of the capillaries?
8 What is the role of tissue fluid?
9 What happens to the excess tissue fluid?

Gaseous exchange in the lungs

Transporting oxygen and carbon dioxide

- Air is breathed into the lungs and enters the <u>alveoli</u>.
- Oxygen in the alveoli is at a <u>higher partial pressure</u> than in the blood.
- Oxygen diffuses from the alveoli into the blood vessels.
- Carbon dioxide in the blood has a higher partial pressure than in the alveoli.
- This causes carbon dioxide to diffuse from the blood into the alveoli in the lungs.

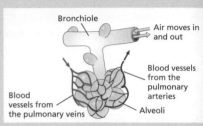

Gas exchange in an alveolus

Air we breathe in contains:

The blood vessels carry blood with low levels of oxygen to the lungs and take away blood high in oxygen to the tissues.

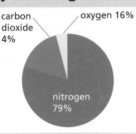

Air we breathe out contains:

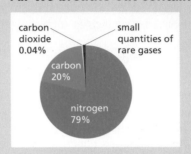

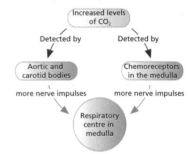

Blood supply to the alveoli

> **EXAMINER'S TOP TIP**
> Control of breathing by monitoring carbon dioxide levels is an example of negative feedback.

How is breathing rate controlled?

The amount of oxygen delivered to the lungs is determined by how quickly we breathe.

However, it is the level of carbon dioxide in the blood that is responsible for the rate of our breathing. High levels of carbon dioxide are monitored by **chemoreceptors** in the **medulla**, and the **aortic and carotid bodies**. The medulla sends impulses along **sympathetic nerves** to the **diaphragm** and **intercostal muscles** to increase the breathing rate.

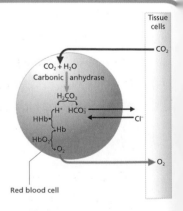

Breathing control mechanism

How haemoglobin works

Haemoglobin absorbs oxygen when the concentration of oxygen is high, as at the lungs. It then releases oxygen when the concentration of oxygen is lower, as at the tissues.

This is what happens at the tissues.

1 **Carbon dioxide** diffuses from tissue into red blood cells.
2 Carbon dioxide combines with water to form **carbonic acid (H_2CO_3)**.
3 This is a reversible reaction using the enzyme **carbonic anhydrase**.
4 Carbonic acid dissociates into **hydrogen ions (H^+)** and **hydrogen carbonate ions (HCO_3^-)**.
5 Hydrogen carbonate ions diffuse out of red blood cells.
6 **Chloride ions** diffuse in to maintain electro-neutrality. This is called the **chloride shift**.
7 Hydrogen ions left in red blood cells reduce the haemoglobin.
8 **Reduced haemoglobin** releases **oxygen**.
9 Oxygen diffuses out of red blood cells and enters tissues.

The process is reversed at the lungs so that haemoglobin picks up oxygen and releases carbon dioxide.

HINT At the lungs just draw the arrows with the arrow-head at the other end and change the label for tissue, to lungs.

The Bohr effect

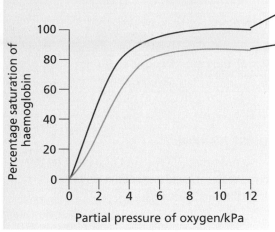

curve showing how haemoglobin combines with oxygen in the lungs

curve showing how haemoglobin combines with oxygen in the tissues

How haemoglobin combines with oxygen

- This is called the **oxygen dissociation curve**.
- It shows that haemoglobin has a high affinity for oxygen.
- An increase in CO_2 causes the curve to move down and right. This causes a release of oxygen for the tissues.
- This means that haemoglobin will always release oxygen at the place and time where it is needed.

Different types of haemoglobin

Foetal haemoglobin is different from normal haemoglobin. It has a higher affinity for oxygen. This ensures that the mother's haemoglobin will always release oxygen for the foetus.

The graph shows the curve for foetal haemoglobin is shifted to the left.

Myoglobin is found in muscles. It has a curve shifted to the left of haemoglobin. This ensures that myoglobin will always take oxygen away from haemoglobin.

Mammals like whales, that spend long periods of time underwater without breathing, have large reserves of myoglobin. It stores oxygen for them.

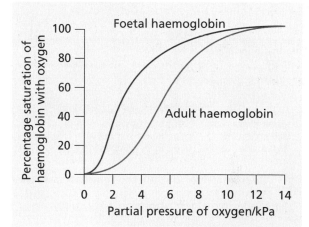

EXAMINER'S TOP TIP
Haemoglobin curves to the left tend to grab hold of oxygen. Haemoglobin curves to the right tend to release oxygen more easily.

Quick test

1 From the oxygen that we breathe in, how much do we absorb into the blood?
2 What part of the brain controls breathing?
3 What substance are chemoreceptors in the carotid body sensitive to?
4 Which nerve transmits the impulse to increase the rate of breathing?
5 What enzyme controls the conversion of carbon dioxide and water to carbonic acid?
6 What does carbonic acid ionise to produce?
7 What do hydrogen ions do to oxyhaemoglobin?
8 What happens to the oxygen dissociation curve when haemoglobin reaches the tissues?
9 Where, on a graph to show oxygen dissociation, is the curve for myoglobin and foetal haemoglobin compared to normal haemoglobin?
10 What is the effect of shifting the oxygen dissociation curve to the left?

1. 4% (20% in, 16% out; 4% of 20% into blood) 2. medulla oblongata 3. carbon dioxide 4. sympathetic 5. carbonic anhydrase 6. hydrogen ions and hydrogen carbonate ions 7. reduce haemoglobin to release oxygen 8. shifts right and down and releases oxygen 9. to the left of normal haemoglobin 10. Haemoglobin acquires a greater affinity for oxygen.

Water transport in plants

- Plants take in water and other substances through their roots.
- Water enters the root hairs by <u>osmosis</u>.
- Once it has entered the root, it travels to the <u>xylem vessels</u>. It can do this by the <u>symplast pathway</u> or the <u>apoplast pathway</u>.

The symplast pathway

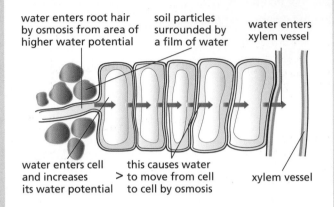

water enters root hair by osmosis from area of higher water potential

soil particles surrounded by a film of water

water enters xylem vessel

water enters cell and increases its water potential

> this causes water to move from cell to cell by osmosis

xylem vessel

Water entering root via the symplast pathway

The apoplast pathway

Water can also pass through the root of a plant by the apoplast route. Water moves between cells rather than passes into them by osmosis.

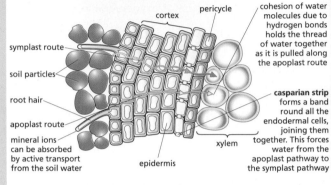

cortex

pericycle

cohesion of water molecules due to hydrogen bonds holds the thread of water together as it is pulled along the apoplast route

symplast route

soil particles

root hair

apoplast route

mineral ions can be absorbed by active transport from the soil water

casparian strip forms a band round all the endodermal cells, joining them together. This forces water from the apoplast pathway to the symplast pathway

xylem

epidermis

Water entering root by the apoplast pathway

- Mineral ions are absorbed by <u>active transport</u>.

Water enters the xylem

- Xylem consists of dead hollow vessels. Each one is like a straw.
- Xylem vessels are strengthened with lignin which stops them collapsing. The picture shows different patterns produced by the lignin.
- The vessels have pits or holes that enable water to pass sideways from one vessel to another.

Xylem vessels

Is it pulled or is it pushed?

Once water has entered the xylem, it has to be carried up to the top of the plant. Some trees are over 100 m in height.

Root pressure ✗
A cut stem exudes water. This is the result of root pressure. But it is NOT the way water gets to the top of the plant.

Osmosis ✗
The maximum height that water can reach by osmosis is about 10 metres. This is NOT the way either.

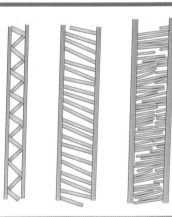

max height 10 metres

Capillarity ✗
This will get the water only a few mm up the stem.

capillarity

Cohesion-tension theory

Water goes up the stem because it is pulled from above. Imagine a thread of water going from the leaf, down the xylem and into the root. As water molecules evaporate from the underneath surface of the leaf the thread of water is pulled up. The thread does not break because the water molecules are held to each other by cohesion.

Demonstrating cohesion-tension theory

This experiment was carried out by Joseph Bohm in 1893. It demonstrates the cohesion-tension theory.

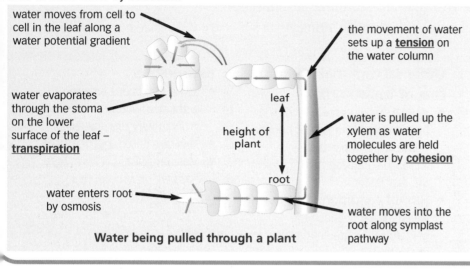

water moves from cell to cell in the leaf along a water potential gradient

water evaporates through the stoma on the lower surface of the leaf – **transpiration**

the movement of water sets up a **tension** on the water column

leaf

height of plant

water is pulled up the xylem as water molecules are held together by **cohesion**

root

water enters root by osmosis

water moves into the root along symplast pathway

Water being pulled through a plant

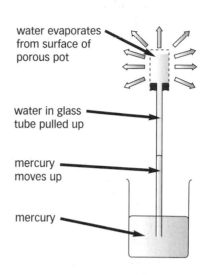

water evaporates from surface of porous pot

water in glass tube pulled up

mercury moves up

mercury

How light affects the rate of transpiration

Plants use light for photosynthesis. They also need carbon dioxide for photosynthesis. Unlike light, carbon dioxide cannot pass through the leaf's surface. Carbon dioxide enters through open **stomata**. This means stomata have to be open during the hours of daylight. The stomata are opened and closed by **guard cells**.

- Potassium (K^+) ions enter the guard cells.
- Malate is produced from starch.
- Water enters by osmosis.
- Guard cells swell and open stomata.

However, at night plants cannot photosynthesise. To stop water loss by transpiration through the stomata, they close them at night.

- Malate is converted to starch.
- K^+ ions leave the guard cells.
- Water leaves by osmosis.
- Guard cells collapse and close the stomata.
- Transpiration stops.

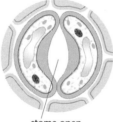

stoma open

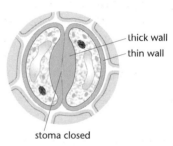

thick wall

thin wall

stoma closed

Quick test

1 Describe the symplast pathway for water to enter plants.
2 Describe the apoplast pathway.
3 Why must water pass through the endodermis by the symplast pathway?
4 Describe xylem vessels.
5 What theory describes how water reaches the top of high trees?
6 Why and how do guard cells close the stomata at night?

1. Water enters root hairs by osmosis and passes from cell to cell along a water potential gradient. 2. Water is pulled along around the outside of cells; the water molecules are held by the hydrogen bonds. 3. The casparian strip stops water moving by the apoplast pathway. 4. dead, hollow lignified vessels 5. cohesion-tension theory 6. To stop water loss. Malate is converted to starch; K+ ions leave the guard cells; water follows by osmosis; guard cells collapse.

Transpiration in plants

Transpiration

<u>Transpiration</u> is the loss of water from any part of the plant. Most of the water is lost by <u>evaporation</u> through the stomata.

- Water molecules evaporate into the sub-stomatal air chamber.
- Water molecules diffuse out through stomata along a concentration gradient.

Measuring transpiration rate

- The rate of transpiration can be measured using a potometer.
- As water evaporates from the leaf, the air bubble moves towards the plant.
- The distance it moves can be measured.
- The bubble can be re-set by using the syringe.
- The apparatus can be used to measure the transpiration rate under different conditions.

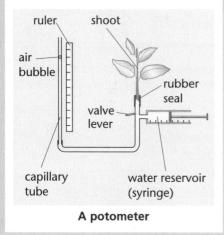

A potometer

Other factors that affect the rate of transpiration

- **Warm air** increases transpiration. The kinetic energy of the water molecules is increased, causing them to leave the leaf faster.
- **Humid conditions** slow down transpiration because the concentration gradient is decreased.
- **Wind** increases transpiration. Water molecules are blown away, maintaining the concentration gradient.

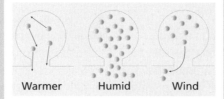

Warmer   Humid   Wind

Xerophytes

- Some plants are adapted to withstand water loss and can grow in very dry places.
- They are called **xerophytes**.
- Marram grass is a xerophyte. It grows on sand dunes.

Marram grass

It has:

- a thick waxy cuticle to reduce evaporation
- fewer stomata
- stomata in sunken pits to build up water vapour
- hairs to reduce air movement around the stomata
- hinge cells that roll up the leaf to trap water vapour
- deep roots to collect water.

How plants transport food

Plants transport sugars through tissue called **phloem**. This can be shown using **radioactive isotopes**.

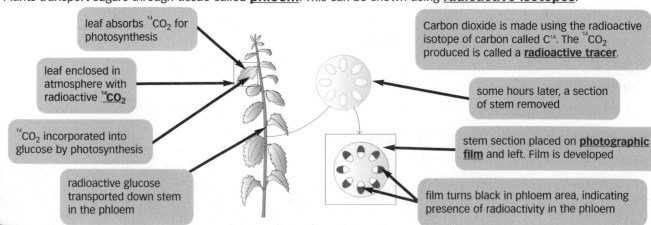

leaf absorbs $^{14}CO_2$ for photosynthesis

leaf enclosed in atmosphere with radioactive $^{14}CO_2$

$^{14}CO_2$ incorporated into glucose by photosynthesis

radioactive glucose transported down stem in the phloem

Carbon dioxide is made using the radioactive isotope of carbon called C^{14}. The $^{14}CO_2$ produced is called a **radioactive tracer**.

some hours later, a section of stem removed

stem section placed on **photographic film** and left. Film is developed

film turns black in phloem area, indicating presence of radioactivity in the phloem

Using radioactive isotopes to show the movement of sugars in a plant

Phloem

- Unlike xylem, **phloem** is living tissue.
- The ends of the phloem cells are connected to each other by **sieve plates**.
- Each **sieve tube** is filled with cytoplasm which can pass to the next cell through the holes in the sieve plate.
- There are no nuclei in phloem cells.
- **Companion cells**, next to the phloem cells, do contain nuclei. These cells perform some metabolic activities for the phloem cell.
- Unlike xylem, phloem can transport substances in both directions. It transports all around the plant.

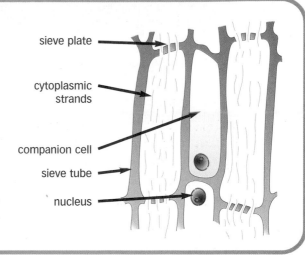

Mass flow hypothesis

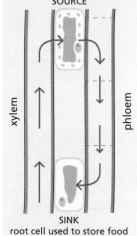

Movement of sugars round a plant in mass flow hypothesis

The **mass flow hypothesis** is an attempt to explain how phloem moves sugars around the plant.
- **Source cells** in the leaf build up sugars.
- The sugars are then actively transported into the phloem.
- The sugars are then transported to the **sink**. The sink can be anywhere in the plant, but is often the root.
- In the sink the sugars are either used for respiration or converted into starch for storage.

How it works
- The **source** is high in sugar.
- Water moves in by **osmosis**.
- This draws water up the xylem.
- This causes pressure in the source.
- Sugar solution is forced down the phloem to the **sink**.
- Sugar is converted to **starch** in the sink.
- Starch is insoluble so the water potential is high in the sink.
- Therefore no water enters the sink by osmosis.

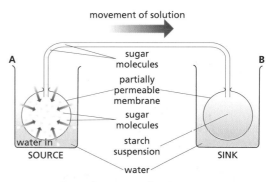

Experiment to demonstrate mass flow hypothesis

Quick test

1 Name the group of plants that are adapted to survive in dry environments.
2 List the ways that marram grass is adapted to survive in a dry environment.
3 List three environmental factors that speed up the rate of transpiration.
4 Explain why transpiration increases in windy conditions.
5 Name a radioactive tracer that can be used to show movement of sugars in phloem tissue.
6 Explain how phloem differs from xylem.
7 Name the theory that is used to explain how transport in the phloem works.
8 Describe how that theory is thought to work.

1. xerophytes 2. thick cuticle, few and sunken stomata, hairs to trap moisture and reduce air flow, hinge cells to roll up leaf, deep roots 3. warmth, dryness, wind 4. Wind blows away water molecules that leave the stomata, thus maintaining a high concentration gradient. 5. $^{14}CO_2$ 6. Sieve tubes are living, have no nucleus and can transport sugars both up and down the plant. 7. mass flow hypothesis 8. Sugar produced in source cells in the leaf is actively transported into phloem. Water follows sugar by osmosis. Increase in pressure forces the sugar solution along phloem to the sink. Here sugar is converted into starch for storage. Osmotic effect then ceases because starch is insoluble. The process is continually repeated as sugar is made in source cells in the leaf.

Exam-style questions

Exchange and transport

1 a List four requirements of a
respiratory surface. [4]

....
....
....

b The diagram is a cross section through a
leaf. Describe how the leaf is adapted for
gaseous exchange. [3]

....
....
....
....
....

2 Counter current flow is the method used by a fish to absorb oxygen into the bloodstream
from the water.
Use the diagram to explain how counter current flow is more efficient. [2]

....
....

| → | 100 | 90 | 80 | 70 | 60 | 50 | 40 | 30 | 20 | 10 | water |
| ← | | 90 | 80 | 70 | 60 | 50 | 40 | 30 | 20 | 10 | 0 | blood |

3 Describe how an oxygen molecule passes from the air in an alveolus into a blood capillary. [2]

....
....

4 Describe the process of inhaling air into the lungs. [4]

....
....

5 Copy the diagram of the heart and on separate
paper add the following labels: [4]
a sino-atrial node
b atrio-ventricular node
c bundle of His
d Purkinje fibres

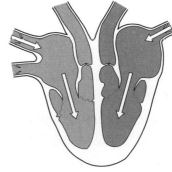

6 Explain the role of the pacemaker. [1]

....

7 Explain the following terms:
a diastole [1]

....

b systole [1]

....

8 The diagram shows the cardiac cycle of a human heart.

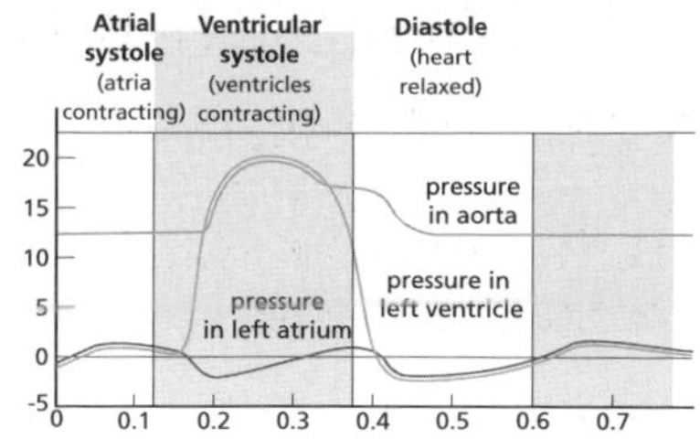

a Calculate the length of the cardiac cycle. [1]

....

b State the maximum pressure during ventricular systole. [1]

....

c State the relationship between pressure and volume
in the chambers of the heart. [2]

....

....

9 The diagram shows the movement of tissue fluid into and
out of a capillary.
Use the diagram to explain this movement. [2]

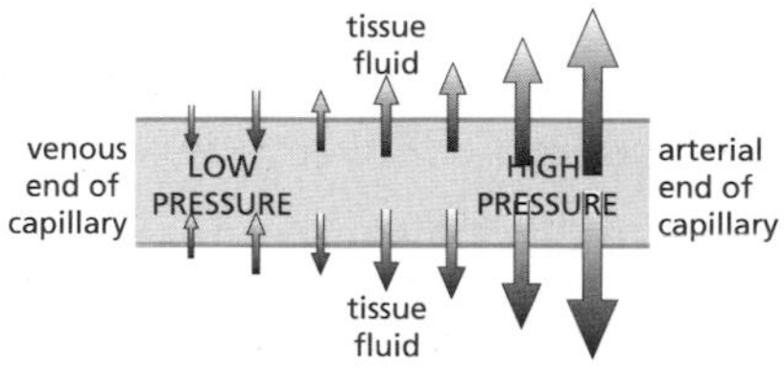

....

....

10 The diagram shows a red blood cell in a capillary next to
actively respiring muscle cells.
State the chemical compounds represented by each of
the numbers. [4]

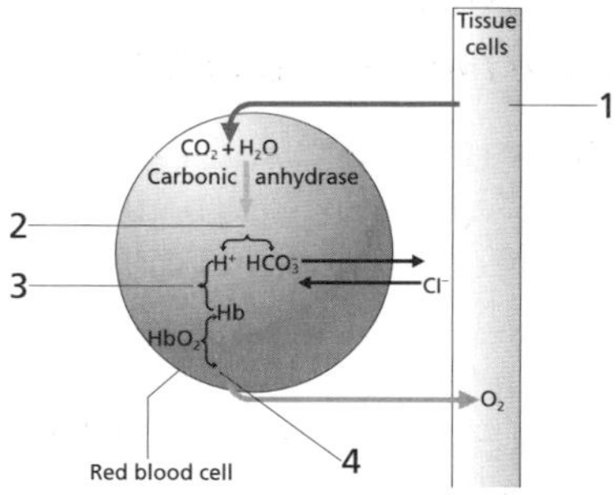

....

....

11 Copy the following graph and draw a line to show the
position of foetal haemoglobin. [2]

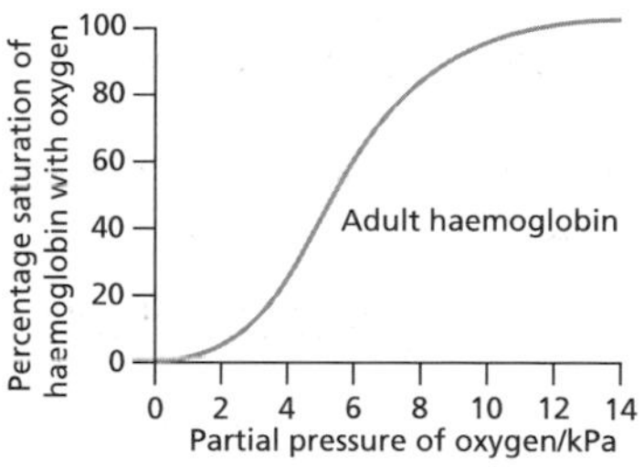

....

....

....

....

12 Describe the difference between the apoplast and symplast route taken by water when it
enters a root. [2]

....

....

13 Describe how the xerophyte in the diagram is adapted to
live in a dry habitat. [3]

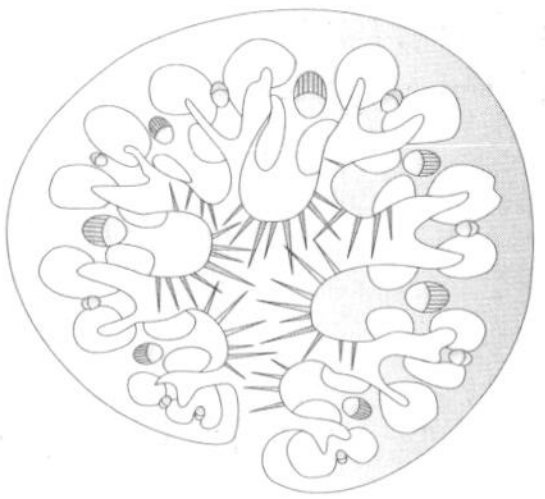

....

....

14 Use the following diagram to explain how the mass flow
hypothesis can be used to explain transport in plants. [5]

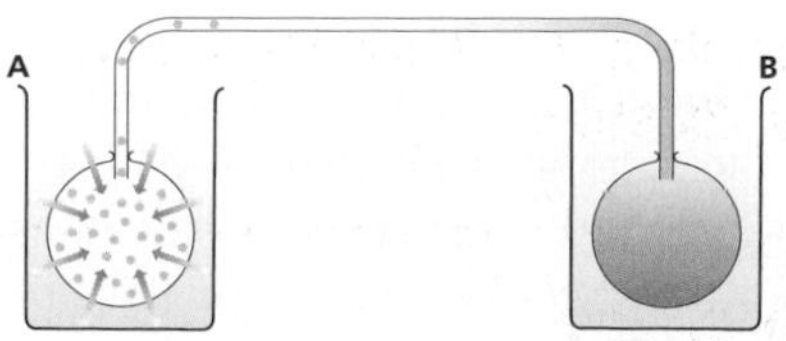

....

....

....

....

....

Total: /44

Mitosis and asexual reproduction

The difference between sexual and asexual reproduction

Asexual reproduction
- **The offspring are all identical – <u>clones</u>.**
- **It can be used to quickly colonise a new environment**
- **It cannot be used for evolution as there is no variation.**
- **The type of cell division involved is called <u>mitosis</u>.**

Sexual reproduction
- **The offspring are all different – there is <u>variation</u>.**
- **It involves male and female.**
- **It enables evolution to occur.**
- **The type of cell division involved is called <u>meiosis</u>.**

Mitosis

- Cells grow and multiply by mitosis.
- The cells that are produced are exact copies of each other.
- They are genetically identical.

- Sometimes some of these cells are used to produce new individuals.
- These new individuals will be genetically identical to their parents. They are **<u>clones</u>**.

EXAMINER'S TOP TIP
Make sure you can relate the pictures to the drawings. You may need to do this in the examination.

Interphase

cell membrane   nuclear membrane
cell wall
nucleolus   chromatin
nucleus

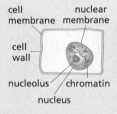

Most of the time the **<u>DNA</u>** in a nucleus is dispersed and the **<u>chromosomes</u>** are not visible. It is now, however, that the DNA makes a copy of itself.

How mitosis works

Prophase

chromosomes condensing and staining darker
nuclear membrane disappears
nucleolus disappearing

When the cell is ready to divide, the DNA shortens and thickens to form visible chromosomes. Each chromosome consists of two **<u>chromatids</u>**. The chromatids are the two copies of the DNA made during interphase.

Metaphase
The **<u>spindle</u>** forms. The chromosomes line up on the equator of the spindle. The chromatids are attached by their **<u>centromeres</u>**.

chromosomes arranged on equator of spindle, held by centromeres
spindle fibres
pole
each chromosome consists of two chromatids

Anaphase
The centromeres divide into two. The spindle fibres shorten and pull the chromatids apart. Because the chromatids are identical it does not matter which way each of them goes.

spindle fibres attached to centromeres
pole
pole
chromatids separate and move to poles
pole

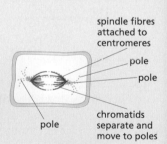

Telophase
A new cell plate forms between the separate chromatids. As the two new cells form, the DNA disperses in order to copy itself and begin the cycle all over again.

chromosomes dispersing
nuclear membrane reforming
spindle disintegrating
cell plate forming

Making clones

All cells contain a full set of genetic information. Some cells are able to use this genetic information to make a whole new organism. Cells that can do this are called **totipotent**.

Not many animal cells are totipotent. But lots of plant cells are. This is why it is much easier to clone a whole plant than a whole animal.

Cloning plants using tissue culture (micropropagation)

This is a form of asexual reproduction or vegetative propagation.

Tissue is removed from a growing plant. **Meristems** (places of active cell division), such as buds, are best.	The tissue is sterilised in very dilute bleach to kill any microorganisms.	The tissue is then blended to separate the cells. Some will be damaged but many are not.	Cells are then placed on a nutrient agar mixture.	Cells grow to form a mass of tissue (**callus**) which differentiates into tiny plants.

The advantage with this method is that one plant can make thousands of clones very quickly. This is ideal for market gardeners who develop a new variety. Normal methods of propagation would take years to produce sufficient plants to sell at a profit.

Embryo splitting in animals

Embryo splitting in animals

Cloning in animals is often done by **embryo splitting**.

- The egg is fertilised by the sperm.
- The fertilised egg divides by mitosis.
- After a few divisions, groups of cells are separated.
- Each new group of cells grows to produce a complete animal.
- This can happen because embryo cells are totipotent.
- This process can occur naturally in humans to produce identical twins.

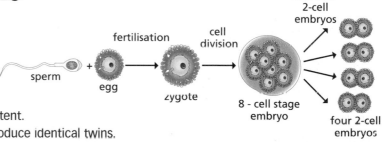

Quick test

1 What type of cell division produces identical cells?

2 What are organisms called that are genetically identical?

3 What is the advantage of sexual reproduction?

4 What happens to DNA during interphase?

5 What are chromatids?

6 What joins two chromatids in a chromosome?

7 What happens during anaphase?

8 What are the advantages of using micropropagation to grow plant clones?

9 How can animal clones be made?

10 What is produced when embryo splitting occurs naturally in humans?

1. mitosis 2. clones 3. produces variation to allow evolution to occur. 4. It is copied or replicated. 5. genetically identical strands of DNA making one whole chromosome 6. centromere 7. The chromatids separate. 8. very quick and cheap 9. embryo splitting 10. identical twins

Meiosis and sexual reproduction

Sexual reproduction involves the fusion of male and female <u>gametes</u>.

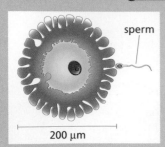

sperm

200 µm

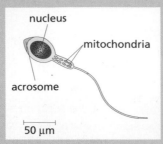

nucleus

mitochondria

acrosome

50 µm

The best way for a sperm to find the ovum is for the ovum to stay still and have large numbers of motile sperm searching for the egg.

Ovum
- **The ovum is much larger than the sperm.**
- **Few ova are produced.**
- **They are incapable of movement.**
- **The ovum contains yolk as food store.**

Sperm
- **Sperm are much smaller than the ovum.**
- **Millions are produced.**
- **Sperm can move to find the ovum.**
- **Mitochondria in the tail provide energy for swimming.**
- **The <u>acrosome</u> helps the sperm to penetrate the ovum.**

Chromosomes and gametes

Normal human body cells contain **46 chromosomes**. If gametes also contained 46 chromosomes, when the ovum was fertilised, the embryo would have 92.
If this doubling happened every generation, the cells would have to contain thousands of chromosomes. There is not room in the nucleus for this number of chromosomes.

The solution to this problem is that gametes have half the number of chromosomes. But this does not mean that half of the genetic instructions will be missing from the gamete … Although we have 46 chromosomes it is much better to think of them as **23 pairs of chromosomes**. We get one set of 23 from our mother and the other set of 23 from our father. Each set of 23 chromosomes is a complete set of the instructions that are needed to make a new human being.

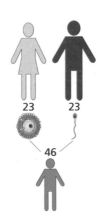

23   23

46

This means because we get two sets, one from mum and one from dad, each nucleus contains two complete sets of genetic instructions.
This is why, when we make our own gametes (sperm or ova), they only need to have 23 chromosomes in them.
Normal cells with 23 pairs of chromosomes are called **diploid**.
Diploid number (2n) = 46
Sperm and ova with only one set of 23 chromosomes are called **haploid**.
Haploid number (n) = 23

Meiosis

How do we get sperm and ova with only half the normal number of chromosomes? During meiosis the cell divides twice. This results in four gametes being produced.

In the cells that make sperm and ova, the chromosomes come together in their pairs.

One chromosome from each pair goes into each new cell. Either chromosome from each pair could go into either cell. This helps to produce variation in the offspring.

The cells with half the chromosome number (haploid) develop into gametes (sperm or ova).

This process is called **meiosis**. You will learn about it in more detail when you do 'A2'.

Interpreting life cycles

Many different organisms use both mitosis and meiosis. However, they often use these two types of cell division at different points in their life cycle.

You need to be able to identify and interpret different kinds of life cycles.

The easiest to start with is the **human life cycle**.

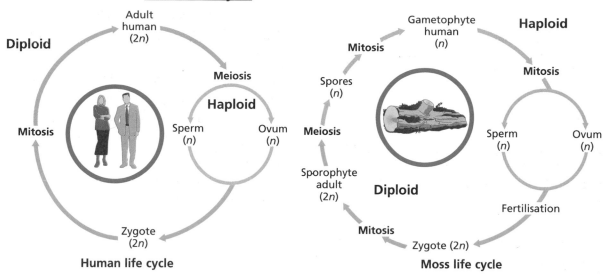

Human life cycle

Moss life cycle

Compare the moss life cycle with the human one. They are very different. The moss has two different kinds of adult. One generation type alternates with the next. It is only the **sporophyte** generation that we normally see. The **gametophyte** generation is very tiny.

Look at the life cycle of the fern. Don't worry about eggs and sperm being there. Surprisingly, many plants use eggs and sperm. See if you can work out which labels are haploid (n) and which are diploid (2n).

EXAMINER'S TOP TIP
A typical examination question would give you a diagram of a life cycle with some labels missing. The questions would then ask you to explain what happened at different stages.

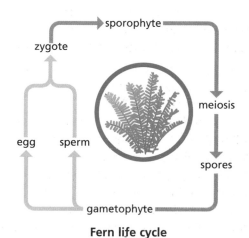

Fern life cycle

HINT Remember that meiosis reduces the number of chromosomes to half.

Quick test

1 List the differences between sperm and ova.

2 What is the number of chromosomes in a normal body cell called?

3 What is the number of chromosomes in a gamete called?

4 How many complete sets of genetic instructions does a normal body cell have and where do they come from?

5 What is the importance of meiosis in reproduction?

6 Where does meiosis occur in the human life cycle?

7 Where does meiosis occur in the life cycle of moss?

1. Sperm are smaller, they move and there are many more of them.   2. diploid   3. haploid   4. two – one set from the mother and one from the father   5. It causes variation as each gamete is different. It reduces the number of chromosomes from 23 pairs to 23 single chromosomes.   6. In gamete formation   7. in spore formation

Sexual reproduction in plants

Many flowers contain both male and female reproductive organs. They are <u>hermaphrodites</u>.

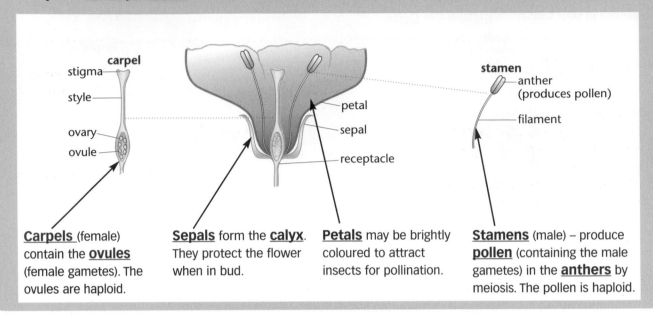

<u>Carpels</u> (female) contain the **<u>ovules</u>** (female gametes). The ovules are haploid.

<u>Sepals</u> form the <u>calyx</u>. They protect the flower when in bud.

Petals may be brightly coloured to attract insects for pollination.

<u>Stamens</u> (male) – produce <u>pollen</u> (containing the male gametes) in the **<u>anthers</u>** by meiosis. The pollen is haploid.

How are ovules and pollen grains made?

<u>Pollen grains</u> are made when a **<u>pollen mother cell</u>** divides by meiosis. In meiosis there are two divisions, so four haploid pollen grains are formed.
Each pollen grain contains two nuclei. One is called the **<u>tube nucleus</u>**, the other is called the **<u>generative nucleus</u>**.

WARNING: Pollen grains themselves are NOT the male gametes. Each pollen grain contains a **<u>tube nucleus</u>** which will form the male gametes.

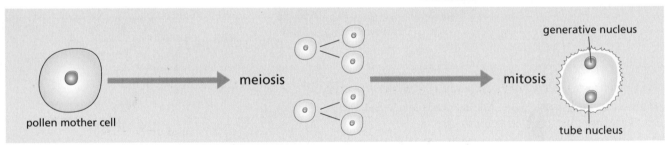

Formation of a pollen grain

<u>Ovules</u> are made when an **<u>embryo sac mother cell</u>** divides to form four haploid cells. Three of these four cells degenerate. (This happens in many other species including humans.) The haploid nucleus inside the remaining cell then begins to divide by mitosis. It divides three times.
This produces a cell containing 8 haploid nuclei. It is called an **<u>embryosac</u>**.

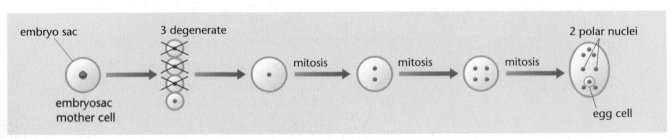

Formation of an ovule

Pollination

Pollination is the transfer of pollen grains from the anther of one flower to the stigma of another. Pollination can be assisted by wind or insects.

EXAMINER'S TOP TIP
Many students get pollination and fertilisation confused. Make sure you know which is which.

Fertilisation

- The pollen grain has landed on the stigma.
- A pollen tube starts to grow out of a gap in the outer coat of the pollen grain and down the style.
- DNA in the tube nucleus codes for the proteins that make the tube.
- The tube nucleus then degenerates.
- The generative nucleus divides by mitosis to make two male nuclei.
- The pollen tube enters the embryosac.
- One male nucleus fuses with the egg cell to make the **zygote**.
- The other male nucleus fuses with the two polar nuclei to form a **triple fusion nucleus**. This will then divide to form the food store for the seed.

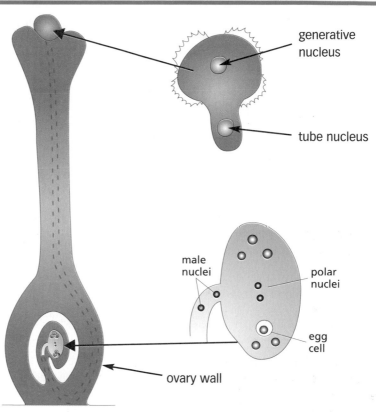

Fertilisation in a flowering plant

What happens after fertilisation?

- The flower dies and the petals fall away. The stigma and style also wither away, leaving the fertilised cells inside the ovary.
- The embryosac will develop into the seed.
- The ovary wall or **pericarp** will develop into the fruit wall.

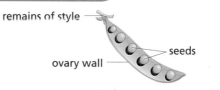

Quick test

1. Name the male parts of the flower.
2. Name the female parts of the flower.
3. When does meiosis take place in the formation of a pollen grain?
4. What happens to the two nuclei in a pollen grain?
5. How is the triple fusion nucleus formed?
6. How many nuclei are found in an embryosac?
7. What does the triple fusion nucleus develop into?
8. What does the fruit develop from?
9. Is the egg cell haploid or diploid?
10. What is the difference between pollination and fertilisation?

1. Stamen (anther, filament)   2. Carpel (stigma, style, ovary)   3. when the pollen mother cell divides to form 4 cells   4. The pollen tube nucleus produces a pollen tube then degenerates; the generative nucleus divides to form two male nuclei   5. by fusion of the two polar nuclei and one male nucleus   6. 8   7. food store for the seed   8. ovary wall   9. haploid   10. Pollination is transfer of pollen from one flower's anther to another flower's stigma; fertilisation involves the fusion of the male nucleus with the egg cell and the fusion of male nucleus with polar nuclei.

Human reproduction

- The organs of the reproductive system are present in the newborn baby.
- As the individual reaches puberty, the sex organs mature and start to function.

Male reproductive system

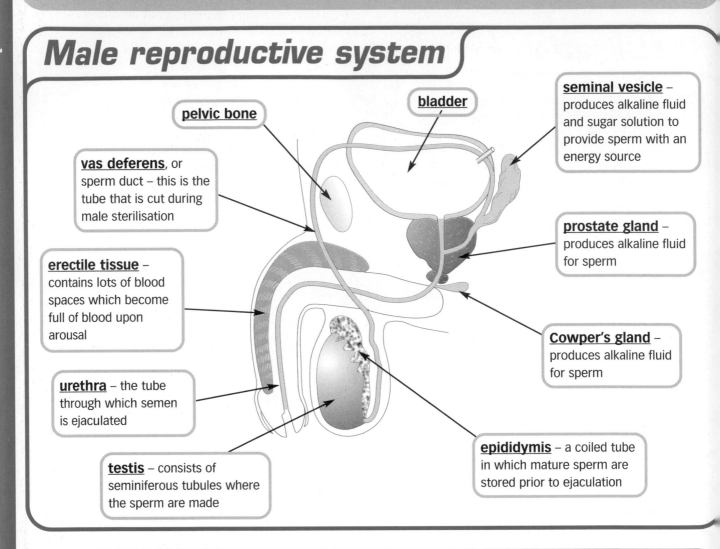

pelvic bone

bladder

vas deferens, or sperm duct – this is the tube that is cut during male sterilisation

erectile tissue – contains lots of blood spaces which become full of blood upon arousal

urethra – the tube through which semen is ejaculated

testis – consists of seminiferous tubules where the sperm are made

seminal vesicle – produces alkaline fluid and sugar solution to provide sperm with an energy source

prostate gland – produces alkaline fluid for sperm

Cowper's gland – produces alkaline fluid for sperm

epididymis – a coiled tube in which mature sperm are stored prior to ejaculation

How sperm are made

The manufacture of sperm is called **spermatogenesis**.
- Inside the testes, the **seminiferous tubules** are lined with **germinal epithelial cells**.
- Germinal epithelial cells divide by mitosis to produce a sperm mother cell.
- The sperm mother cell divides by mitosis to produce many more sperm mother cells.
- These sperm mother cells are often called **spermatogonia**.
- The sperm mother cells develop into **primary spermatocytes**.
- Primary spermatocytes divide by meiosis to form four spermatids.
- Spermatids are immature sperm. They develop into mature sperm by being fed by special cells in the seminiferous tubules.
- Mature sperm are then stored in the epididymis and vas deferens, ready to be used.

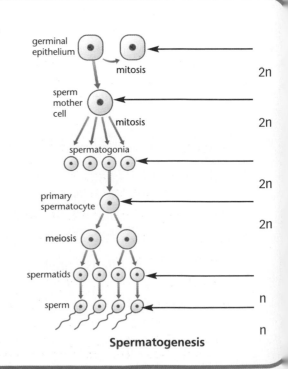

germinal epithelium — mitosis — $2n$

sperm mother cell — mitosis — $2n$

spermatogonia — $2n$

primary spermatocyte — $2n$

meiosis

spermatids — n

sperm — n

Spermatogenesis

Female reproductive system

ovaries – in which the female egg cells, or ova, develop

fallopian tube – if fertilisation is going to happen, it takes place here. Ciliated cells help to move the ovum towards the uterus.

uterus – in which any fertilised ovum will implant. The thick muscular walls (**endometrium**) have a good blood supply to nourish the developing foetus.

cervix

vagina

How ova are made

Ova are made by a process called **oogenesis**. It may look similar to spermatogenesis but there are important differences.

- Germinal epithelial cells divide by mitosis to produce **ovum mother cells** or **oogonia**.
- Each oogonium divides by mitosis to produce many more oogonia.
- The oogonia then develop into **primary oocytes**.
- Each primary oocyte is surrounded by other cells to form a **primary follicle**. At this point, development is arrested.
- All the follicles are diploid and they are formed before the female baby is born. At birth, thousands of primary follicles are already present, ready to be used later.
- Starting during adolescence, a primary follicle matures each month.
- It divides by meiosis to form four cells. One develops into a **secondary oocyte**; the other three are called **polar bodies** and they degenerate.
- The secondary oocyte or **ovum** is released from the surface of the ovary. The empty follicle develops into the **corpus luteum**.

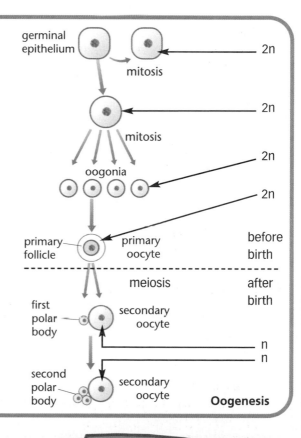

germinal epithelium — 2n
mitosis
— 2n
mitosis
oogonia — 2n
— 2n
primary follicle — primary oocyte — before birth
---------- meiosis — after birth
first polar body — secondary oocyte
— n
— n
second polar body — secondary oocyte
Oogenesis

Quick test

1 **What produces a sugar solution that the sperm use as a food source?**
2 **Where are mature sperm stored?**
3 **State two differences between spermatogenesis and oogenesis.**
4 **When does meiosis take place in spermatogenesis and oogenesis?**
5 **How does the ovum pass down the fallopian tube?**
6 **Where does fertilisation take place?**

1. seminal vesicles 2. epididymis 3. A primary oocyte has arrested development and produces only one secondary oocyte.
4. before the production of the spermatids and before the production of the secondary oocyte 5. ciliated cells that line the fallopian tube cause the ovum to move towards the uterus 6. in the fallopian tube

The menstrual cycle and fertilisation

The female menstrual cycle is controlled by hormones under the influence of the pituitary gland.

The role of hormones

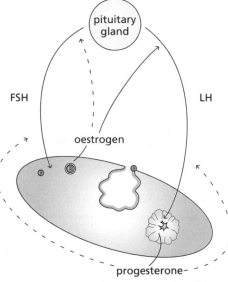

- The pituitary gland releases **follicle stimulating hormone** (**FSH**).
- FSH stimulates the growth of primary follicles in the ovary. One of them develops into a mature **Graafian follicle** containing a secondary oocyte.
- The secondary oocyte starts to produce the hormone **oestrogen**.
- The oestrogen inhibits the production of FSH, but stimulates the production of **luteinising hormone** (**LH**).
- **Ovulation** takes place on day 14 – the secondary oocyte (the ovum) is released from the Graafian follicle.
- The ruptured follicle becomes the **corpus luteum** which produces the hormone **progesterone**.
- Progesterone inhibits the production of both FSH and LH.
- As FSH and LH levels fall, this in turn causes the corpus luteum to stop producing progesterone.
- Progesterone levels fall and this triggers the start of menstruation.
- The fall in progesterone causes the pituitary gland to start producing FSH and the whole cycle then repeats itself, taking approximately 28 days to complete each time.

Hormones in the menstrual cycle

The mentrual cycle

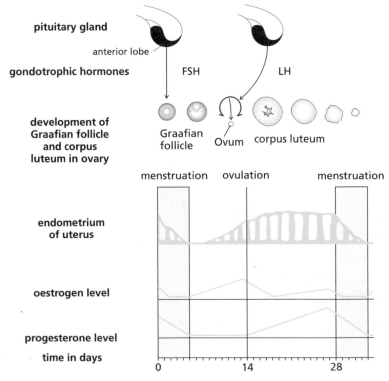

The stages of the menstrual cycle

- A Graafian follicle matures, releases an ovum and then turns into the corpus luteum that produces progesterone.

- The endometrium thickens ready to receive the fertilised ovum, but breaks down producing menstrual flow if it is not fertilised.

- Oestrogen level peaks before ovulation.

- The fall in progesterone acts as a trigger to menstruation.

EXAMINER'S TOP TIP
Try to relate what is happening on the graphs to what is happening to the ovary and follicle.

Fertilisation

The sperm are deposited in the vagina. From there, they have to swim to get into the fallopian tubes in order to fertilise the ovum.

The head of the sperm has a structure called the **acrosome**. This contains enzymes that digest the protective layer surrounding the ovum, allowing the head of the sperm to enter.

The male nucleus can then fuse with the female nucleus and the **zygote** will be diploid (2n).

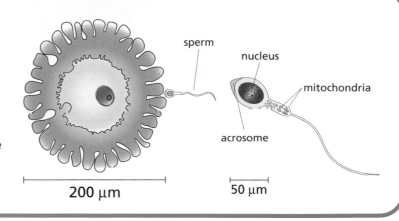

200 μm   50 μm

Hormones in pregnancy

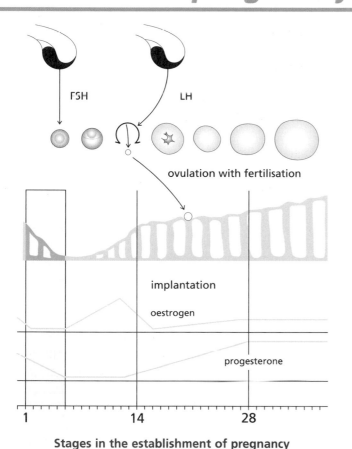

FSH   LH

ovulation with fertilisation

implantation

oestrogen

progesterone

1   14   28

Stages in the establishment of pregnancy

- The pituitary gland will eventually secrete **prolactin** which stimulates milk production for breastfeeding.

- The corpus luteum continues to develop and produce progesterone.

- Sperm fertilises the ovum in the fallopian tube.

- After implantation, the endometrium continues to thicken. A **placenta** is formed to supply the developing embryo with nutrients and to remove waste.

- Oestrogen level is maintained.

- Progesterone level remains high and prevents menstruation from starting.

Quick test

1 What effect does FSH have on the ovary?

2 What effect does oestrogen have on FSH?

3 What effect does oestrogen have on LH?

4 What stimulates the production of progesterone?

5 What happens when progesterone levels fall?

6 What happens to the levels of progesterone if pregnancy occurs?

7 Which gland produces prolactin?

8 What does prolactin do?

1. stimulates development of the Graafian follicles   2. inhibits   3. stimulates   4. the action of LH on the corpus luteum   5. menstruation starts   6. levels remain high   7. pituitary   8. stimulates milk production

Pregnancy

- **Pregnancy takes place when the sperm fertilises the ovum inside the fallopian tube.**

From ovulation to implantation

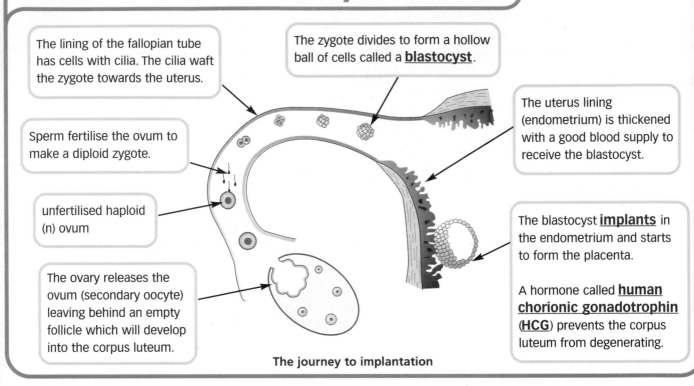

The lining of the fallopian tube has cells with cilia. The cilia waft the zygote towards the uterus.

The zygote divides to form a hollow ball of cells called a **blastocyst**.

The uterus lining (endometrium) is thickened with a good blood supply to receive the blastocyst.

Sperm fertilise the ovum to make a diploid zygote.

unfertilised haploid (n) ovum

The ovary releases the ovum (secondary oocyte) leaving behind an empty follicle which will develop into the corpus luteum.

The blastocyst **implants** in the endometrium and starts to form the placenta.

A hormone called **human chorionic gonadotrophin** (**HCG**) prevents the corpus luteum from degenerating.

The journey to implantation

The role of the placenta

In the placenta, the foetal blood vessels and the mother's blood vessels lie very close to each other, allowing the exchange of oxygen, carbon dioxide, food and waste substances.

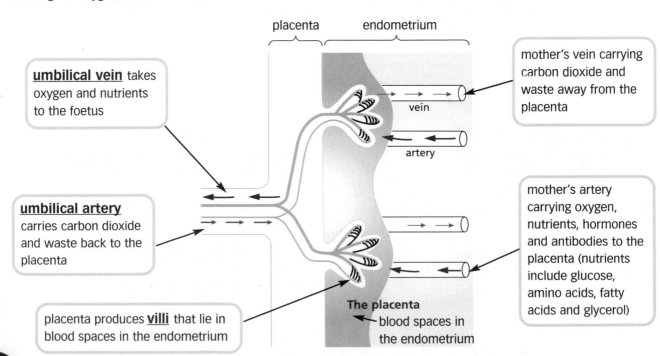

placenta   endometrium

umbilical vein takes oxygen and nutrients to the foetus

mother's vein carrying carbon dioxide and waste away from the placenta

vein

artery

umbilical artery carries carbon dioxide and waste back to the placenta

mother's artery carrying oxygen, nutrients, hormones and antibodies to the placenta (nutrients include glucose, amino acids, fatty acids and glycerol)

placenta produces **villi** that lie in blood spaces in the endometrium

The placenta
blood spaces in the endometrium

Birth

Approximately 40 weeks after fertilisation, progesterone levels fall which causes birth to take place. The powerful uterine muscles contract, stimulated by the hormone **oxytocin**, and push the baby out.

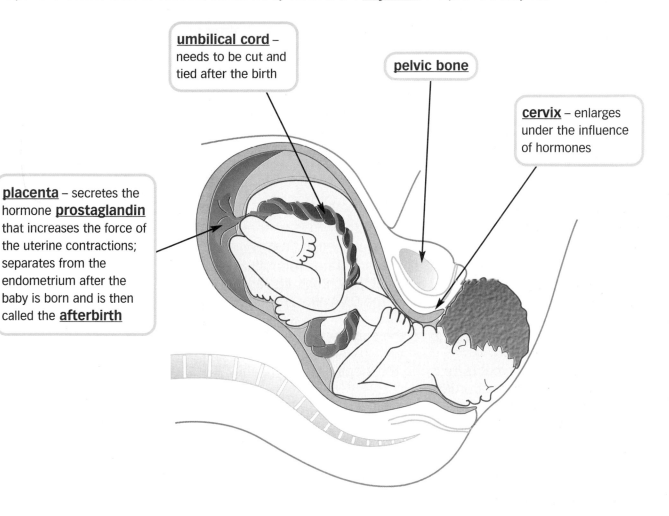

umbilical cord – needs to be cut and tied after the birth

pelvic bone

cervix – enlarges under the influence of hormones

placenta – secretes the hormone **prostaglandin** that increases the force of the uterine contractions; separates from the endometrium after the baby is born and is then called the **afterbirth**

The fall in the level of progesterone stimulates the production of the hormone **prolactin**, which in turn stimulates **lactation** – the production of milk. The first milk, **colostrum**, is very rich in the mother's antibodies. Colostrum ensures that the baby is protected from disease during the first few weeks until it builds up its own defence mechanisms.

Quick test

1 *Where does fertilisation take place?*

2 *What is the cell produced by the fusion of sperm and ova called?*

3 *What is the hollow ball of cells called that is produced when the zygote starts to divide?*

4 *Where does the ball of cells implant?*

5 *Is the blastocyst diploid or haploid?*

6 *Which hormone does the blastocyst produce?*

7 *What does this hormone do?*

8 *List three substances that pass from the mother's blood into the umbilical vein*

9 *What surrounds the villi of the placenta?*

10 *Which hormone is secreted by the placenta prior to birth?*

11 *What does this hormone do?*

1. fallopian tube 2. zygote 3. blastocyst 4. the endometrium in the uterus 5. diploid 6. human chorionic gonadotrophin 7. stops the corpus luteum from degenerating 8. glucose, oxygen, hormones, amino acids, fatty acids, glycerol 9. open blood spaces 10. prostaglandin 11. increases strength of uterine contractions

Exam-style questions

Reproduction

1 The picture shows a cell dividing.

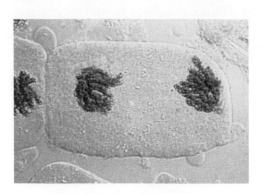

a Name the stage of the dividing cell.. [1]

b Put the following statements about cell division in the correct order.   [3]
  i   New cell plate forms between the separate chromatids.
  ii The spindle fibres shorten and pull the chromatids apart.
  iii When the cell is ready to divide, the DNA shortens and thickens to form visible
      chromosomes.
  iv The chromosomes line up on the equator of the spindle.

2 The following statements describe the steps for cloning animal embryos.
   Put them in the correct order.   [4]
   i   After a few division groups of cells are separated.
   ii The egg is fertilised by the sperm.
   iii This can happen because embryo cells are totipotent.
   iv Each new group of cells grows to produce a complete animal.
   v   The fertilised egg divides by mitosis.

3 The drawing shows a cell about to divide by meiosis.
   a State which of the diagrams, a, b, c or d, shows the chromosomes that would be
   found in a
   gamete.

[1]

a

b

c

d

b Explain the reason for your choice.   [2]

....

....

4 The diagram shows the life cycle of a moss plant.

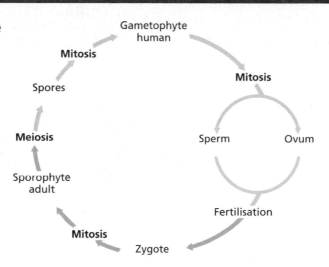

a Label each stage to show whether it is haploid or diploid. [2]

b Show where meiosis takes place. [1]

5 The following statements describe stages in the fertilisation of a flowering plant. Put them in the correct order. [7]

  i A pollen tube is produced which starts to grow out of a gap in the outer coat, and down the style. ☐

  ii DNA in the tube nucleus codes for the proteins that make the tube. ☐

  iii The other male nucleus fuses with the two polar nuclei to form a triple fusion nucleus. This will then divide to form the food store for the seed. ☐

  iv The tube nucleus then degenerates. ☐

  v The pollen grain lands on the stigma. ☐

  vi The generative nucleus divides by mitosis to make two male nuclei. ☐

  vii The pollen tube enters the embryosac. ☐

  viii One male nucleus fuses with the egg cell to make the zygote. ☐

6 The graph shows the hormone levels of oestrogen and progesterone in the blood of a woman.

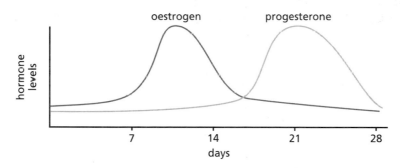

a State whether or not the woman is pregnant and explain the reason. [1]

....

b Explain how the follicle stimulating hormone (FSH) would have affected the hormone oestrogen on day 7, as shown in the graph. [1]

....

c State where on the graph you would expect ovulation to have taken place. [1]

....

Total: /24

DNA

Where is DNA found?

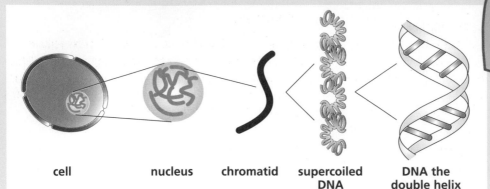

cell   nucleus   chromatid   supercoiled DNA   DNA the double helix

Genes

Genes are sections of **DNA**. A single gene consists of a length of DNA that codes for one specific protein. This is sometimes referred to as 'one gene – one protein – one job'.

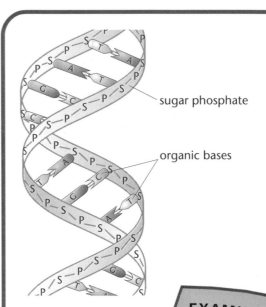

sugar phosphate

organic bases

DNA – the double helix

What is DNA?

DNA is often referred to as a **polynucleotide**. This is because it is made up from single units called **nucleotides**.

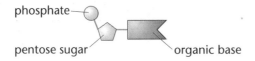

phosphate

pentose sugar   organic base

A single nucleotide

A nucleotide is made up from one of four bases, attached to a pentose sugar (it has 5 carbon atoms), and a phosphate.

There are four different bases: **adenine**, **thymine**, **cytosine** and **guanine** (**A**, **T**, **C** and **G**). These four bases are the four letters of the genetic alphabet. The nucleotides are joined together in pairs. The base of each pair is joined by a hydrogen bond.

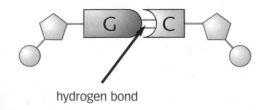

hydrogen bond

DNA

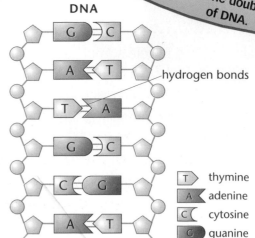

hydrogen bonds

T	thymine
A	adenine
C	cytosine
G	guanine

What does DNA do?

DNA has two different jobs.

1 It provides the **code** for all the instructions needed to make a living organism.

The code consists of a language using only four letters. Using only combinations of four letters (bases), the code has to be very long. A single instruction to make a protein may have several thousand letters. Each human being has enough DNA to stretch to the moon.
Because DNA is double stranded, only one strand is the actual code. It is called the **sense strand**.

2 It has to copy itself (**replication**) so that at cell division all new cells have a copy of the complete set of instructions.

Every time a cell divides, each new cell must receive a copy of the DNA. This means that during a lifetime DNA gets copied thousands of times. Whenever errors occur in the structure of the DNA, they are copied when the DNA is replicated. These errors are called **mutations**.

How does DNA replicate?

For a long time, scientists were not sure how DNA replicated.
Did it make a complete copy of itself, in one go? – the **conservative method**.

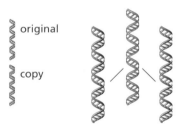

Or did it make a copy of just one single strand each time the cell divided? – the **semi-conservative method**.
Eventually they discovered it was the semi-conservative method.

The semi-conservative method of replication

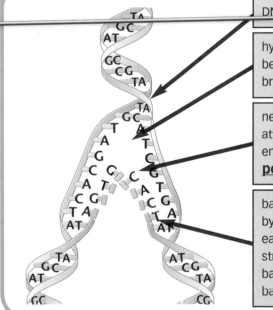

DNA spiral untwists

hydrogen bonds between bases are broken

new bases are attached by the enzyme **DNA polymerase**

bases held together by hydrogen bonds – each new double strand has 50% new bases and 50% old bases

Quick test

1 What is the relationship between a chromatid and a double strand of DNA?
2 What name is given to a section of DNA that codes for one protein?
3 How many bases are in one nucleotide?
4 Which base pairs with adenine?
5 Which base pairs with guanine?
6 What are the two jobs of DNA?
7 What holds base pairs together?
8 What is the name of the strand of DNA that carries the code?
9 Which method does DNA use for replication, conservative or semi-conservative?
10 How much of the DNA is copied in the semi-conservative method?
11 How much of the DNA is original DNA, in a replicated section of DNA?

1. They are the same thing. 2. a gene 3. one 4. thymine 5. cytosine 6. coding for information, replication 7. hydrogen bonds 8. sense strand 9. semi-conservative 10. 50% 11. 50%

The DNA code

How DNA codes for proteins

Proteins consist of about 20 different amino acids. This means that, unlike the English language that has thousands of different words, the language of DNA has only about 20 different 'words' that it has to code for.

How the code works

EXAMINER'S TOP TIP
Some amino acids have more than one three-letter code. Other three-letter codes can be used to indicate the end or start of a gene.

Imagine that in the language of DNA, one letter (base) meant one 'word'.

letters	words
A	– A
T	– T
C	– C
G	– G

This would give only four words which is not enough for 20 amino acids.

What if the language of DNA used two letters to make a 'word'?

letters	words			
A	AA	AC	CT	TG
T	TT	AT	CG	GA
C	CC	AG	TA	GC
G	GG	CA	TC	GT

This gives sixteen words. It is better, but still not enough for 20 amino acids.

What if the language of DNA used three letters to make a 'word'?

letters	words
A	AAA
T	AGC
C	TAG
G	CGG

64 different combinations

This gives 64 possible 'words' – more than enough with some to spare, for the 20 amino acids.

How DNA makes proteins

We now know that three bases are needed to code for one amino acid. We also know that proteins are made by linking amino acids together to form long chains. But what are the steps in between?

Step 1
The DNA double helix unzips for the whole of the section that codes for a particular protein.
This happens because the hydrogen bonds between the bases are broken.

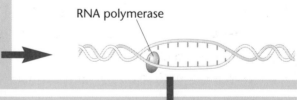

Step 2
An enzyme called **RNA polymerase** makes a single-strand copy of the sense strand that codes for the protein. This is called **transcription**. However, the copy is slightly different. The base **thymine** is replaced by a new base called **uracil**. The strand is called **RNA**, not DNA, and because it carries the coded message it is called **messenger RNA** (**mRNA**). Notice how the U and A now pair together rather than T and A.

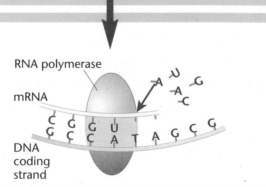

Step 3

The single strand of mRNA leaves the nucleus through a **nuclear pore**. Three bases on the mRNA are called a **codon** and together they code for one specific amino acid. So if you want to know how many bases code for a particular protein, it is the number of amino acids multiplied by three.

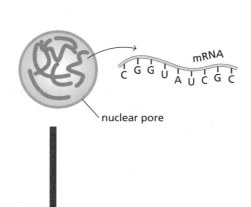

nuclear pore

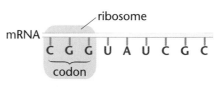

mRNA

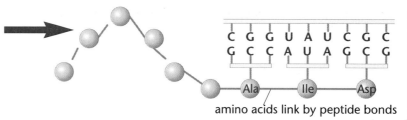

ribosome

mRNA

codon

EXAMINER'S TOP TIP
Codons are sometimes referred to as <u>triplets</u> because they contain three bases.

Step 4

The mRNA lines up on a ribosome. Remember, ribosomes are the cell organelles where proteins are made.

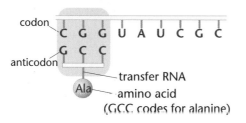

codon
anticodon
transfer RNA
amino acid
Ala
(GCC codes for alanine)

Another type of RNA called transfer RNA (**tRNA**) has a matching **anticodon** to which is attached a specific amino acid. Each different anticodon has a different amino acid. The anticodon in the picture, GCC, codes for the amino acid called alanine. This is called **translation**.

Step 5

Each codon is read in turn. An anticodon attaches a new amino acid. The amino acids bond together by a condensation reaction to form peptide bonds. Finally the newly formed protein peels away and folds into its secondary and tertiary structure.

C G G U A U C G C
G C C A U A G C G
Ala   Ile   Asp

amino acids link by peptide bonds

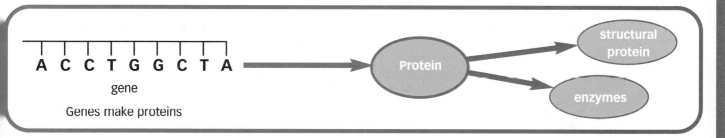

A C C T G G C T A
gene
Genes make proteins

Protein

structural protein

enzymes

Quick test

1 How many bases are needed to code for one amino acid?
2 Why are pairs of bases not used to code for one amino acid?
3 What enzyme is used to make mRNA?
4 What is the copying of DNA into mRNA called?
5 What base is absent in RNA?
6 What replaces this base?
7 When mRNA leaves the nucleus, where does it go?
8 What is the name for a group of three bases on mRNA?
9 Which RNA transfers amino acids to the mRNA?
10 What is this process called?
11 What are the three bases on tRNA called?

1. three 2. Two bases could code for only 16 amino acids and 20 combinations are needed. 3. RNA polymerase 4. transcription 5. thymine 6. uracil 7. the ribosome 8. codon 9. tRNA 10. translation 11. anticodon

Mutation and gene technology

A mutation to DNA simply means that the DNA has a spelling mistake!
Just as in the English language, spelling mistakes sometimes do not matter too much and it is possible to work out what the sentence means.

My aunt boght a new coat.
(My aunt bought a new coat.)

Sometimes the spelling mistake is very important and changes the meaning of the whole message.

My aunt bought a new goat.

In DNA, 'spelling mistakes' can happen when one of the bases gets changed. This gives rise to a <u>mutation</u>. There are various ways that this can happen. Some are more serious than others.

Types of mutation

Substitution
One of the bases gets replaced by a different base. The triplet CTT in the DNA for normal haemoglobin can get changed to CAT. The change from thymine to adenine results in sickle cell anaemia.

Inversion
Part of the DNA gets turned round in the opposite direction.

G A T T C G C A T G C A T A A C T G C T C   before
G A T C T G C A T G C A T A A C T G C T C   after

This means that the second codon, TCG, now reads CTG and codes for a different amino acid. This could change the structure of the protein.

Insertion or deletion
This is when a new base is added or an existing base is removed.

G A T T C G C A T G C A T A A C T G C T C   before
G A T T C G A T G C A T A A C T G C T C   after

The base cytosine is completely removed. This deletion changes all of the triplets that occur after it. It is a very serious mutation and completely changes the protein. Insertions are equally as damaging.

Translocation
Whole sections of DNA are moved to somewhere else. The degree of damage depends upon which section is translocated.

What effect does a mutation have?

Small changes in the base sequence of DNA will code for different amino acids.

This means the protein will be a different shape.

However some mutations can be useful. We are the result of many useful mutations. But useful mutations are very rare.

Different amino acids mean that the sulphur bridges that produce the secondary and tertiary structures will be in a different position.

If the protein is an enzyme, and the change in shape affects the active site, the enzyme will no longer work.

This in turn can cause a metabolic block to occur in a metabolic pathway.

Gene technology

Many modern genetic processes require large amounts of DNA. The **polymerase chain reaction** is a method of producing large amounts of DNA from a single DNA molecule. For example, a scene of crime investigation may find only small quantities of DNA. The procedure can be used to provide larger quantities to carry out genetic fingerprinting tests.

The DNA is mixed with:
- a solution containing all four bases, A, T, C and G
- the enzyme DNA polymerase
- short lengths of RNA called **primers** – these bind with the single strand and act as a trigger to start the copying process.

1 DNA is heated to 95 °C for 30 seconds. This causes the two strands to separate by breaking the hydrogen bonds between the bases.

2 DNA is rapidly cooled to 37 °C for 30 seconds. This causes primers to bind to complementary strands of DNA.

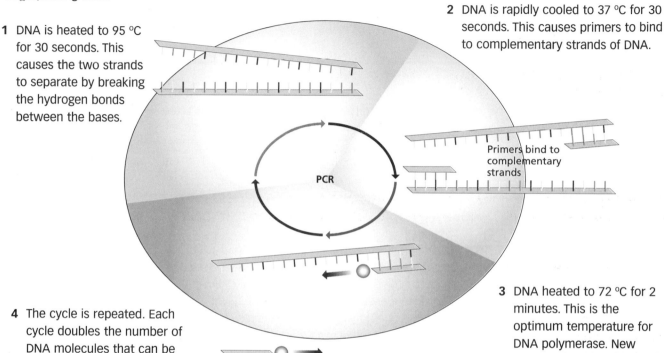

Primers bind to complementary strands

PCR

4 The cycle is repeated. Each cycle doubles the number of DNA molecules that can be used as templates for the next cycle.

3 DNA heated to 72 °C for 2 minutes. This is the optimum temperature for DNA polymerase. New strands of DNA are made.

Polymerase chain reaction

Copying genes

The process is now fully automated using a PCR machine. After only 25 cycles, more than 1 million copies of DNA can be produced from a single strand of DNA.

PCR can be used to provide DNA for electrophoresis and for determining the sequence of the bases in DNA.
The Human Genome Project has now deciphered the whole of the human genome.

Quick test

1 What is a substitution mutation of DNA?

2 What is a deletion mutation of DNA?

3 DNA can sometimes be moved to a different location on the chromosome. What is this called?

4 Which type of mutation can have the most serious consequences?

5 How can an enzyme be affected by a mutation of the DNA?

6 What does PCR stand for?

7 What is the purpose of RNA primers?

8 Why is it called a chain reaction?

9 What other substances are needed for the reaction to take place?

1. One or more bases of DNA is removed and replaced by a different base.   2. A section of DNA is removed.   3. translocation   4. insertion and deletion when the base is added or deleted because it has a knock-on effect to all the DNA strands. All amino acid codes can be changed.   5. A mutation can code for a different amino acid. This will alter the positions of the sulphur bridges that determine the secondary and tertiary structure of the enzyme. This in turn alters the enzyme's shape. If the active site is affected, the enzyme may be useless.   6. polymerase chain reaction   7. They provide a starting point for the reaction.   8. Each cycle doubles the DNA available to act as a template for the next cycle.   9. a supply of the four bases and the enzyme DNA polymerase

Genetic engineering

All living organisms on the planet use DNA as a code to produce life. Unlike human languages used for communication, there is only one language in DNA and it is used by all living organisms. This means that any DNA taken from one organism will be decoded by any other organism. A tortoise will understand the DNA from a rose and vice versa.

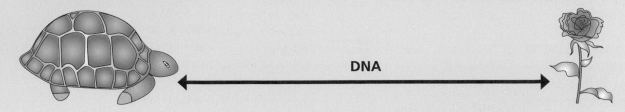

DNA

There does not seem to be much point in putting tortoise DNA into a rose. In some cases, however, putting DNA from one organism into another can be very useful.

The insulin story

Some people cannot make their own <u>insulin</u>, the <u>hormone</u> that <u>controls levels of sugar</u> in the blood. The condition is known as <u>diabetes</u>. Many diabetics can survive only by having daily injections of the insulin. The insulin used to be obtained from animals. This was very expensive and some people became allergic to <u>animal insulin</u>. If they could not use the insulin, they would die. <u>Genetic engineering</u> enables scientists to produce large quantities of <u>human insulin</u> very cheaply.

1 Normal human DNA is extracted, containing the gene for making insulin.

2 The insulin gene is cut out using an enzyme called a **restriction enzyme**.

3 The same restriction enzyme is used to cut open the DNA in a bacterial **plasmid** – a small ring of DNA in addition to the normal DNA strand found in bacteria.

4 The human insulin gene is inserted into the plasmid. The ends of DNA are rejoined using an enzyme called a **ligase**.

5 The bacteria are then allowed to multiply by growing and dividing in a fermenter. Millions of copies of the bacteria are made, all containing the human insulin gene.

6 The gene then starts to produce human insulin which can be harvested.

Genetic markers

Scientists sometimes want to know which bacteria have been successful in taking up the gene. One way they can do this is to use **genetic markers**. Some plasmids contain a gene that makes the bacteria resistant to particular antibiotics. These genes can be used as genetic markers.

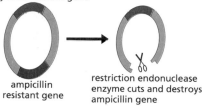

tetracycline resistant gene

ampicillin resistant gene

restriction endonuclease enzyme cuts and destroys ampicillin gene

insulin gene added with ligase enzyme

1 This plasmid has two antibiotic resistance genes, one for tetracycline and one for ampicillin. The insulin gene can be inserted in the middle of the ampicillin gene.

2 The bacteria are then grown on a culture medium in a Petri dish. The bacteria grow to form colonies which are visible to the naked eye. In the picture above, the colonies that contain the insulin gene are

coloured red but in reality the scientists would not know which colonies contained the insulin gene.

3 A piece of filter paper can be placed on top of the first Petri dish and then transferred to a second dish, producing an exact copy of the position of each culture. This time, however, the nutrient agar contains the antibiotic tetracycline. It destroys all those bacteria not resistant to tetracycline. All the insulin-containing bacteria will survive.

4 The same process is repeated onto another Petri dish. This is called **replica plating**. This time the agar contains the antibiotic ampicillin. This will kill all those bacteria that do contain the insulin gene. By comparing plates 2 and 3, the scientists can determine which bacterial colonies contain the human insulin gene.

Large-scale culturing

The **bacteria** that have been identified as containing the insulin gene can then be cultured in a **fermentation vessel** on a large scale to produce marketable quantities of the **human hormone, insulin**.

Many other products, such as antibiotics, hormones and enzymes, can also be made using this process. Even food such as **single-cell protein (SCP)** can be made as a cattle food. Conditions such as temperature, pH and aeration need to be kept constant. The culture may be run continuously, with the product being harvested and new nutrients being fed into the mixture.

The bacteria that have been identified as containing the insulin gene can then be cultured in a fermentation vessel on a large scale to produce marketable quantities of the human hormone, insulin.

Many other products, such as antibiotics, hormones and enzymes, can also be made using this process. Even food such as single-cell protein (SCP) can be made as a cattle food. Conditions such as temperature, pH and aeration need to be kept constant. The culture may be run continuously with the product being harvested and new nutrients being fed into the mixture.

Quick test

1 How many kinds of DNA languages are there?

2 Name the enzyme used to cut DNA apart.

3 Name the enzyme used to stick DNA together.

4 Name the rings of DNA found in bacteria that contain the genes for antibiotic resistance.

5 Name a human hormone that has been produced by genetic engineering.

6 When a bacterium has been genetically engineered, what is the next step in the process?

7 How do scientists know that a particular bacterium contains the new gene?

8 Name two antibiotic-resistant genes that can be used as genetic markers.

9 What is meant by replica plating?

Gene therapy

Genetic disease

Cystic fibrosis is a genetic disease. It is caused by a mutation of base 1627 in the cystic fibrosis gene. The mutation results in a missing amino acid from a protein called the <u>CFTR protein</u>.

Normal CFTR protein

CFTR protein is found in cell membranes. Its job is to transport chloride ions out of the cell, through the cell membrane.

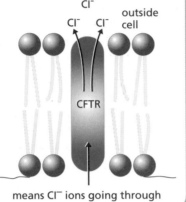

eytoplasm

means Cl⁻ ions going through

Faulty CFTR protein

When the protein is faulty, it can no longer transport chloride ions out of the cell. This causes water to enter the cell from outside the membrane by osmosis. The mucus on the outside of the membrane loses water and thickens.

The thick mucus causes breathing problems and infections. It also blocks the ducts of the digestive glands and reproductive ducts, causing sterility.

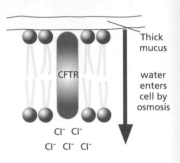

The solution

A cold virus is a microorganism that works by injecting its DNA into the cells of our respiratory tract. The virus DNA instructs our own DNA to stop working and to make more copies of the virus instead. We actually end up making the viruses that are going to make us ill.

Scientists can use viruses to get useful DNA into cells. The virus that does this is called a **<u>vector</u>**.

First the viral DNA is destroyed and the new CFTR gene inserted into the virus. The method used is the same one that is used on bacterial plasmids.

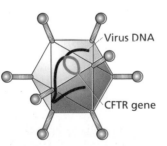

epithelial cell

virus DNA injected into cell

Virus DNA

CFTR gene

Another solution

Another way that the new gene can be placed into the cells is to wrap the gene inside lipid molecules. Because the plasma membrane is a phospholipid sandwich, lipid molecules can pass easily across the membrane.

The new gene can be carried across the membrane by the lipid molecules.

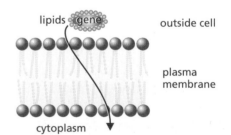

lipids gene — outside cell

plasma membrane

cytoplasm

How gene therapy can help treat cystic fibrosis

Cystic fibrosis is a hard-disease to treat. Every single cell of the person's body carries the faulty genes. It is impossible to repair the DNA of every single cell. However, some cells are more adversely affected by the faulty gene than others.

If the epithelial cells lining the breathing tubes could be repaired, the quality of life would be significantly improved.

This type of treatment is called **gene therapy**. The problem, however, is how to get the new DNA into the cells of the respiratory tract. One method that is being trialled is to use an aerosol containing viral vectors.

Genetically modified animals

Organisms that contain genes from different organisms are called **transgenic**. Some transgenic sheep have been produced that contain the human gene for making **alpha-1 antitrypsin**. This is an enzyme that helps to protect the lungs of humans from damage during an infection. It is very useful to be able to give this enzyme to patients who suffer from cystic fibrosis, or other lung diseases such as emphysema.

How the sheep were genetically modified

The only way to produce transgenic animals is to insert the new gene into the cell that forms the zygote. When the zygote divides, the new gene is copied into every single cell of the organism. The picture shows a micro-needle about to inject new DNA into the zygote of a sheep. The zygote is being held on the end of a micro-pipette.

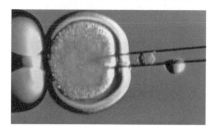

Sheep milk containing alpha-1 antitrypsin

The ethical issues of genetic engineering

Candidates are expected to be able to discuss the **ethical** and **moral** issues involved in **genetic engineering**. Here are some of the questions that arise.

- If we can save lives by curing diseases such as **cystic fibrosis**, should we not use the knowledge?
- Is it right to change the **DNA of crops** such as wheat just to make them **resistant** to a particular kind of weed killer?
- Rice does not contain vitamin A. People who eat mainly rice can go blind because of a lack of the vitamin. A strain of GM rice now exists that contains the vitamin A gene from a carrot. People who eat the rice no longer go blind. Should scientists be allowed to do this?
- Big companies are getting the **patents to any gene** that they discover. Other scientists will have to pay if they wish to use this gene. Is it right that someone should be able to patent our DNA or should the knowledge belong in the public domain?

EXAMINER'S TOP TIP
Remember, there are no right or wrong answers. You just need to show that you are aware of the issues and can debate them from both sides of the argument.

Quick test

1 What is the protein that is faulty in people with cystic fibrosis?

2 What does this protein do in healthy people?

3 What effect does the faulty gene have in people who have cystic fibrosis?

4 Name two methods of transferring a healthy gene into the epithelial cells of a patient with cystic fibrosis.

5 Why does a virus make a good vector for the gene?

6 What are organisms called that have DNA from other organisms?

7 What enzyme can be produced from genetically modified sheep?

8 List two arguments for the use of genetic engineering.

9 List two arguments against the use of genetic engineering.

1. CFTR 2. Transport chloride ions out of the cell through the plasma membrane 3. Chloride ions accumulate inside the cell. This causes water to enter by osmosis. The mucus on the outside of the cell loses water and so thickens and blocks the lungs and other ducts in the body. 4. Virus vector; gene enclosed in lipid molecules. 5. Viruses naturally inject their DNA into cells. This means they can be used to inject genetically modified DNA into cells. 6. Transgenic 7. Alpha-1 – Antitrypsin 8. Producing rice with vitamin A to prevent blindness; curing genetic diseases such as cystic fibrosis 9. Unknown side effects; morality or religious argument

Exam-style questions Use the questions to test your progress. Check your answers on pages 94–95.

Chromosomes, genes and DNA

1 The following diagram is a section of the sense strand from a DNA molecule.

C G C G A T C T G A G A A T G

The table shows the sequence of bases on tRNA that code for specific amino acids.

CUG	leucine
AUG	methionine
AGA	arginine
CGC	arginine
GGC	glycine
GAU	aspartic acid

a How many amino acids does the strand code for? .... [1]

b What does the table tell you about the amino acid arginine? [1]
....

c Write down the sequence of bases on the mRNA molecule. [1]
....

d Write down the sequence of amino acids in the protein molecule that would be made. [1]
....

2 a Describe two roles carried out by DNA. [2]
....
....

b Compare the conservative and semi-conservative hypotheses of DNA replication. [2]
....
....

3 Look at the following strand of DNA before and after it has mutated.

C G C G A T C T G A G A A T G   before mutation

C G C G A T C T G A A A T G   after mutation

a State the name of this type of mutation. .... [1]

b Explain the effect this will have on the structure of the protein. [2]
....
....

c Explain why this type of mutation is usually more serious than inversion. [2]
....
....

4 The diagram shows the steps in the DNA polymerase chain reaction. 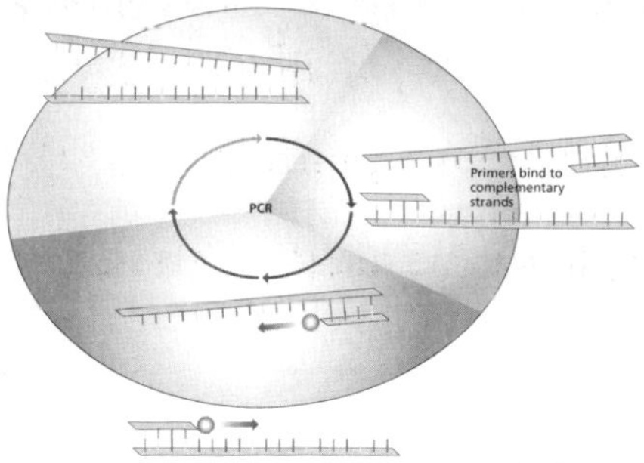
  a Explain why the DNA polymerase chain reaction is used. [1]
  ....
  ....
  ....

Primers bind to complementary strands

PCR

  b Describe the four main steps in the process. [4]
  ....
  ....
  ....

  c What role do primers play in the process [1]
  ....

5 Genetic engineering can involve the transfer of DNA from one organism to another.
  The diagram shows a human insulin gene being inserted into a bacterium. 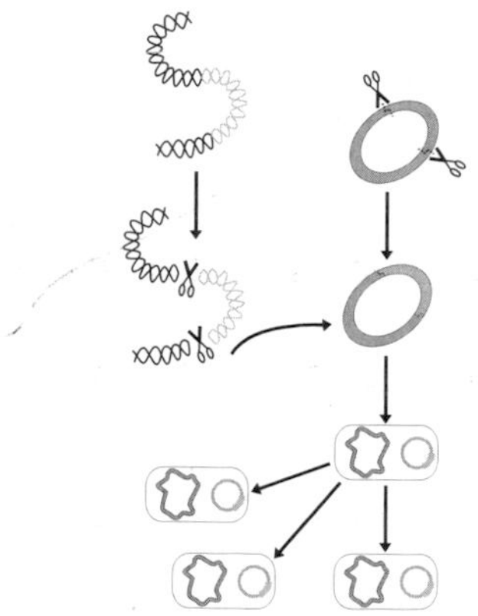
  a Explain how the process works. [5]
  ....
  ....
  ....
  ....

  b Explain why genetic markers are sometimes used. [1]
  ....

6 Cystic fibrosis is a genetic disease.
  a Use the diagram of part of a normal cell plasma membrane, to explain how the symptoms of cystic fibrosis are caused. [3]
  ....
  ....
  ....
  ....

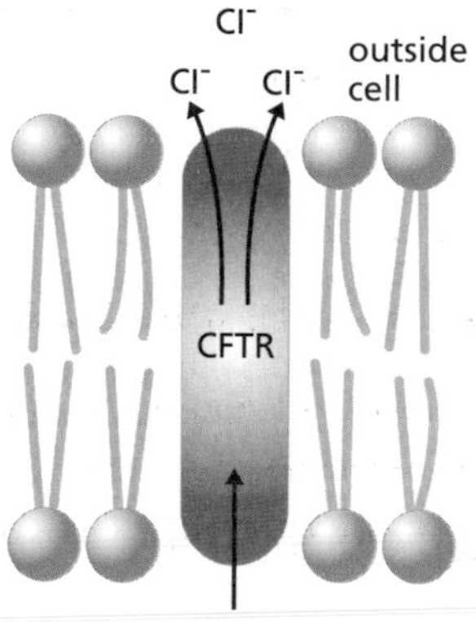

Cl^- outside cell
Cl^- Cl^-
CFTR
Cl^- cytoplasm

  b Explain how genetic engineering using a viral vector may soon be used to cure cystic fibrosis. [3]
  ....
  ....
  ....

  c Describe how genetically engineered sheep can be used to help treat cystic fibrosis. [3]
  ....
  ....

7 Describe two moral arguments, one in support of, and one against, the use of genetic engineering. [2]
  ....
  ....

Total: /38

Digestion

Digestion is the process whereby an organism breaks down large complex food molecules into smaller ones that it can use for energy and growth. The process relies on enzymes.

Extracellular digestion

- Saprophytic fungi digest food using extracellular digestion.
- Enzymes are secreted from the fungus onto the food.
- The enzymes break down the large insoluble food molecules into smaller soluble molecules.
- The smaller molecules are absorbed into the fungus.
- The fungus then uses these molecules for energy and growth.

→ enzymes

⇀ digested food

How to see extracellular digestion

- Fungus mould (*Rhizopus*) is grown on agar jelly in a Petri dish. The agar contains starch.
- The fungus mould grows to the size of a ten pence coin.
- The Petri dish is then flooded with iodine solution.

- The agar turns black but an area around the fungus remains clear. This is because the fungus has released the enzyme carbohydrase. The carbohydrase has digested the starch into simple sugars.

This process can be used to measure how active the enzyme is. The more active the enzyme, the bigger the clear area around the mould. This is called an **assay**.

fungus

black

clear area

fungus

petri dish   agar jelly

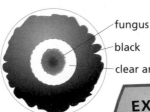

> **EXAMINER'S TOP TIP**
> Learn these enzymes. Remember – endopeptidases break peptide bonds; exopeptidases remove individual amino acids.

Digestion in humans

Location	Secretion	Enzyme	Substrate	Product
mouth	saliva	amylase	starch	maltose
stomach	gastric juice	pepsin (endopeptidase)	protein	polypeptides
duodenum	pancreatic juice	lipase	lipids	fatty acids and glycerol
		more amylase	starch	maltose
		endopeptides	protein	polypeptides
		exopeptidases	polypeptides	amino acids
	bile	none	emulsifies lipids	none
small intestine	intestinal juice	maltase	maltose	glucose

Bile is not an enzyme. It emulsifies lipids so that the enzyme lipase has a larger surface area to work on.

A balanced diet

A balanced diet contains:
- **carbohydrates** – sugars and starch to provide the body with energy
- **protein** – to provide the body with body-building amino acids for growth and repair
- **fats and oils** – these are high in energy and provide insulation from cold; they are also used to make phospholipids
- **fibre** – undigestible cellulose that gives food bulk and stimulates peristalsis in the gut
- **vitamins** – essential chemicals needed in small quantities
- **minerals** – inorganic ions such as calcium for bones and iron for haemoglobin
- **water** – the solvent in which all chemical reactions take place.

Daily requirements

If we eat too much we get fat.
If we eat too little we starve.
How much food do we need?

- baby – 2400 kj
- child – 8900 kj
- teenage male – 12600 kj
- teenage female – 9600 kj
- adult male – 11600 kj
- adult female – 9500 kj

Variations

- male in heavy work – 16600 kj
- female in heavy work – 12600 kj
- pregnant female – 10500 kj
- breastfeeding mother – 11400 kj

The human digestive system

mouth – saliva contains amylase to break down starch to maltose

oesophagus – has thick muscle layer for squeezing food down by peristalsis

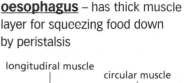
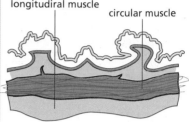

longitudiral muscle   circular muscle

Oesophagus wall

stomach – contains glands that secrete hydrochloric acid and the precursor to pepsin, pepsinogen

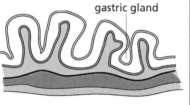

gastric gland

Stomach wall

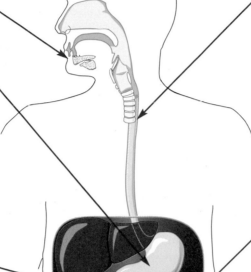

gall bladder – secretes bile to emulsify fat (NB bile is not an enzyme)

large intestine – where water is absorbed

ilium – has lots of villi to increase surface area for absorption

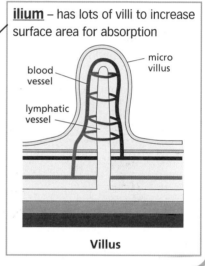

micro villus
blood vessel
lymphatic vessel

Villus

> **EXAMINER'S TOP TIP**
> This is a topic that is often used for analysing data such as graphs, tables and charts.

Quick test

1. *What is digestion that occurs outside of cells called?*
2. *Give an example of an organism that uses this type of digestion.*
3. *How can this type of digestion be used as an assay for enzyme activity?*
4. *Why should bile not be referred to as an enzyme?*
5. *Where is the enzyme pepsin found and what does it do?*
6. *What is fibre and why is it part of a balanced diet?*
7. *Name two minerals that are needed by the body and the function they perform.*
8. *List the five layers of the gut starting with the epithelial layer.*
9. *What is the purpose of villi in the small intestine?*
10. *What is the purpose of digestion?*

1. extracellular   2. fungus mould, Rhizopus   3. using starch agar Petri dish assay.   4. It brings about a physical change in size of the food (fat), and not a chemical change.   5. in the gastric juice in stomach; breaks protein into polypeptides   6. indigestible cellulose; provides bulk, stimulating the muscles of the gut to perform peristalsis   7. calcium for bones and teeth; iron for haemoglobin   8. epithelium, mucosa, submucosa, circular muscle, longitudinal muscle   9. They increase surface area for absorption.   10. It breaks large insoluble molecules down into small soluble ones that can be absorbed into the bloodstream.

Lifestyle and coronary heart disease

Factors that contribute to good health
- a balanced diet
- exercise

Exercise should be carried out about three times a week and last for at least 30 minutes.
Heartbeat should be increased to up to 60% above normal resting rate, provided you do not have a medical condition.
Exercise helps to: relieve stress, reduce resting heartbeat rate, speed recovery and increase stamina and suppleness.

Factors that can damage health
- smoking
- excessive drinking
  Too much alcohol can cause cirrhosis of the liver. The maximum safe amount varies, but for men is about 20 units a week and for women about 16 units a week.
- too much stress
  Stress releases the hormones cortisol and adrenaline. Their purpose is to prepare the body for action in order to escape a dangerous situation – they are designed for short-term use. If released over a long term, the effects on the body can be damaging.

Diet and heart disease

In a balanced diet, the amount of saturated fat should be restricted. Saturated fats are found mainly in animal fat. Foods such as meat, milk, butter and cheese are all high in saturated fats. Eating too many saturated fats can lead to **atherosclerosis**.
- A healthy artery allows the blood to flow freely over the smooth **endothelial lining** of the blood vessel.
- Yellow fat deposits are laid down on the wall of the artery.
- The fatty deposits are made from **cholesterol** which is taken in with our food and is also made in the liver. The deposits build up to form an **atheroma** and bloodflow is restricted. This may, in turn, lead to blood clots.
- Blood pressure increases as the heart finds it harder to pump blood through the blood vessels.
- Eventually the atheroma may block the blood vessel and blood can no longer flow through it.

endothelial

healthy artery

atheroma

clot

atheroma blocks artery

Aneurysm

Increased blood pressure can cause the damaged wall of the blood vessel to balloon outwards. If an aneurysm bursts, it can cause a rapid loss of blood due to internal bleeding, leading to death.

aneurysm

Myocardial infarction

Sometimes an atheroma or blood clot can block an artery that supplies the cardiac muscle with blood.
A partial blockage can cause angina or pain.
A complete blockage causes a heart attack.

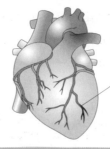

blood vessel gets blocked

Stroke

Blood clots can also block other small blood vessels throughout the body. A blood clot that forms a blockage in the brain causes a **stroke**.

The effect of smoking on health

What do cigarettes contain?

The four main dangerous groups of substances contained in cigarette smoke are:

Carcinogens damage the DNA of the cells in the lungs. This can eventually lead to lung cancer.

Tar coats the lining of the alveoli in the lungs. It slows down gaseous exchange and permanently damages the alveoli.

Nicotine is addictive. It makes people want to continue smoking.

Carbon monoxide combines irreversibly with haemoglobin and prevents it from carrying oxygen.

EXAMINER'S TOP TIP
Make sure you can relate each component of cigarette smoke to its specific effect on the body.

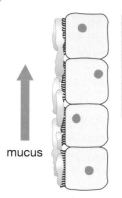

mucus

- When the cigarette smoke enters the lungs, it paralyses and destroys the **cilia** (small hairs) that line the epithelial cells of bronchi and bronchioles. The job of these small cilia is to constantly waft upwards the mucus that forms in the lungs, carrying with it the dirt and bacteria that we have breathed in.

How are the lungs damaged?

- Eventually the walls of the alveoli become damaged and leak tissue fluid. This is called **emphysema**. This increases coughing but also reduces the surface area for gaseous exchange. Advanced cases need oxygen cylinders in order to be able to breathe.

- The mucus eventually reaches the back of the throat and is swallowed or coughed away. When the cilia are paralysed by the cigarette smoke, the mucus is not carried away and builds up. This causes a **smoker's cough**.

damaged alveoli cause emphysema

- Finally the carcinogens in the tar cause permanent damage to the DNA which may cause some cells to divide uncontrollably. This malignant growth is called a **cancer**.

- The build-up of mucus can also lead to an infection of the lungs and bronchi. This is called **bronchitis**.

Quick test

1. **What kind of fat in our diet can lead to atherosclerosis?**
2. **What are the fatty deposits lining the artery walls made from?**
3. **What is the fatty lump called?**
4. **Why is atherosclerosis dangerous?**
5. **What is the name for a weakening and ballooning of the artery wall?**
6. **What will happen if a blood clot blocks part of the cardiac artery?**
7. **What does a partial blockage of the cardiac artery cause?**
8. **When do you think that angina would be most noticeable?**
9. **List four constituents of tobacco smoke.**
10. **What is emphysema?**

1. saturated fats 2. cholesterol 3. atheroma 4. It can block blood vessels, increase blood pressure and cause blood clots to form. 5. aneurysm 6. myocardial infarction (heart attack) 7. angina (chest pain) 8. after exercise 9. carbon monoxide, tar, nicotine, carcinogens 10. reduction in surface area of alveoli; damage to the alveoli causing tissue fluid to leak into the spaces

Disease

Infection
Infection is caused when the body is invaded with a **pathogen**. Pathogens include microorganisms such as **bacteria**, **viruses**, **fungi** and **protozoa**.

Mental diseases
From what has been learned about the brain, we know that **mental illness** is often a result of an **imbalance** in the **brain chemistry** of **neurotransmitters**.

Diet-related diseases
These diseases are caused by **eating too much** or **too little** of certain foods. They include diseases such as **obesity**, **anorexia nervosa** and **vitamin deficiency** diseases such as **scurvy**.

Genetic diseases
These are diseases that are **inherited from our parents**. This means that we are born with them. They are called **congenital diseases**.

Environmentally caused disease
Some diseases are caused by things in our **environment**. They include **lung cancer** from smoking and **skin cancer** from exposure to **ultraviolet rays** in sunlight.

Auto-immune disease
This is when the body's own **defence mechanism** turns against itself. Examples include **asthma** and **rheumatoid arthritis**.

disease

Infectious diseases

Different types of infectious diseases are transmitted in different ways.
Some diseases are transmitted by more than one method.

Droplet infection
Coughs and sneezes spread diseases, especially when droplets carrying the disease float through the air.

Direct contact
Sexual intercourse is the most intimate form of direct contact. Syphilis, AIDS and gonorrhoea are three examples of diseases transmitted in this way.

Contaminated food or water
Sometimes we just put the pathogen straight into our mouths. Cholera in contaminated water supplies and salmonella on chicken are two examples of this type of infection.

Vector
A vector is something that carries a disease. It may be another person with germs on their hands, or a mosquito carrying malaria.

Contaminated blood products and needles
Contaminated blood transfusions have given some patients hepatitis C. Drug users have contracted AIDS from using contaminated hypodermic needles.

Definitions
Endemic – a disease that is always present in a population, e.g. the common cold
Epidemic – an outbreak of a disease in a specific area such as a country
Pandemic – a disease that spreads over a wider area, e.g. crossing continents

Tuberculosis

The <u>tuberculosis</u> (<u>TB</u>) bacteria attack the lungs causing scarring of lung tissue. Bleeding occurs and nodules of bacteria form. As the lung damage continues, breathing becomes more difficult. The infected person loses weight and often contracts other diseases.

Spread

<u>Tuberculosis</u> is spread by <u>droplet infection</u> when someone sneezes or coughs and another person breathes in the droplets containing the <u>bacteria</u>.

Cause

It is caused by a bacterium called *mycobacterium tuberculosis* – rod-shaped bacilli.

Treatment

The usual treatment is a course of several different <u>antibiotics</u> taken for many months. Unfortunately there is now an <u>antibiotic-resistant</u> strain of tuberculosis. This means that antibiotics are ineffective against it, making it impossible to cure anyone who has the disease.

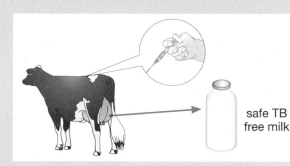

safe TB free milk

Prevention

1 <u>Hygiene</u>. TB is a disease often found in unhygienic overcrowded conditions. Good housing and health care can help to minimise the spread of the disease. However, we should not be complacent – anyone can catch it.

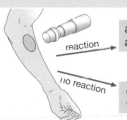

reaction → already has antibodies

no reaction → given BCG vaccination

2 <u>Eradication of TB</u> in cattle. Cattle are now TB tested to make sure they are free of the disease. Also, milk is pasteurised to kill off any harmful bacteria. This is done by heating the milk to 72 °C for 15 seconds and then rapidly cooling it.

3 <u>Heaf test</u> and <u>vaccination</u>. A small amount of tuberculin antigen is injected into the skin. If the skin reddens and swells, antibodies to the disease are already present. If there is no reaction, the person is vaccinated with the BCG injection. This contains attenuated (weakened) tuberculosis bacteria to stimulate the production of antibodies.

4 <u>Mass screening</u>. <u>X-rays</u> can detect the lung damage caused by the bacteria even before the patient is aware of the infection.

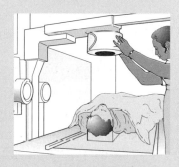

Quick test

1 List six different types of disease.

2 List five ways diseases can be transmitted.

3 What does endemic mean?

4 What does epidemic mean?

5 What does pandemic mean?

6 How is TB spread?

7 What causes TB?

8 How can TB be prevented?

9 What test is used to detect if a person already has the antibodies to TB?

10 How does the Heaf test work?

1. infectious, genetic, environmentally caused, auto-immune, diet-related, mental   2. droplet, direct contact, contaminated food, blood products, vectors.   3. a disease that is always present in a population   4. a disease that spreads through an area such as a country   5. a disease that spreads over a much larger area such as from one continent to another   6. droplet infection   7. Mycobacterium *tuberculosis*   8. better hygiene, TB-free cattle and milk, vaccination, mass screening   9. the Heaf test   10. Small amounts of tuberculin antibody are injected into the skin. If the person is immune, their antibodies will cause a reaction indicated by reddening and swelling of the skin.

AIDS

Acquired immune deficiency syndrome (AIDS) is caused by the HIV virus. HIV stands for human immunodeficiency virus. It is a retrovirus, containing RNA, not DNA.

Spread
The disease is spread by direct intimate contact of body fluids. The virus is easily destroyed once outside the body. This makes it impossible to catch the virus unless a transfer of fluid takes place.

Intimate contact
Virus particles in semen or vaginal fluid can be transferred during sexual intercourse.

Sharing needles
The virus left on the needle can then be injected into another person.

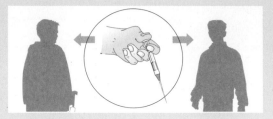

Infected blood products
Blood is now tested and treated to ensure it is HIV-free.

HIV virus in the human body

1 HIV virus – contains **two strands of RNA** and the **enzyme reverse transcriptase**.

2 The protein coat of the virus attaches itself to the cell membrane of a human **T-lymphocyte**.

3 RNA strands and reverse transcriptase enzyme enter the lymphocyte.

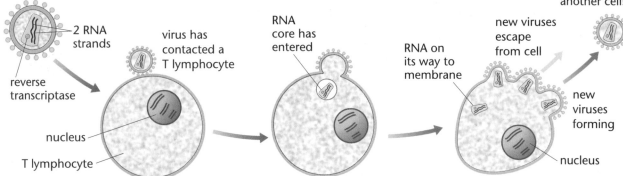

HIV virus

2 RNA strands

reverse transcriptase

nucleus

T lymphocyte

virus has contacted a T lymphocyte

RNA core has entered

RNA on its way to membrane

new viruses escape from cell

ready to infect another cell!

new viruses forming

nucleus

4 The reverse transcriptase causes the cell to produce DNA copies of the viral RNA.

5 The new DNA enters the nucleus of the lymphocyte.

6 The DNA may remain dormant for years. Once activated, it instructs the lymphocyte's DNA to make more copies of the HIV virus.

7 Viruses are then released to invade other T-lymphocytes.

8 As more and more lymphocytes are destroyed, the body can no longer fight invading pathogens. The person then has AIDS.

Treatment
- There is no known cure at present.
- Drugs can be used only to slow down the progress of the virus.
- The best hope lies in new advances in the science of genetic engineering.

Prevention
Prevention is the key in the fight against AIDS. The spread of the HIV virus can be prevented by:
- use of condoms for 'safe sex'
- remaining faithful to one sexual partner
- discarding and destroying used needles
- screening blood products.

Cholera

Cholera causes extreme vomiting and diarrhoea. The body rapidly loses fluids and ions through the gut wall and unless these are replaced rapidly, death occurs.

It is a disease that is much more likely to occur during times of war, famine or extreme poverty. It has been completely eradicated from the UK but would soon return if standards of hygiene deteriorated.

Spread
Cholera is a water-borne disease. It occurs when drinking water becomes contaminated with human faeces.

Cause
- Cholera is caused by a microorganism called a **vibrio**. It is a comma-shaped bacterium called *Vibrio cholerae*.

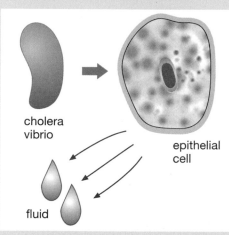

cholera vibrio

epithelial cell

fluid

- It releases powerful toxins that attack the epithelial lining of the gut.
- The cells produce an enzyme called **adenyl cyclase**.
- This causes fluid to be secreted into the gut resulting in severe diarrhoea.
- The vibrio lives in water and can survive long periods of time outside the human body.

Treatment and prevention

- Antibiotics such as **tetracycline** can be used to kill the bacteria.
- Fluid and salts have to be replaced either by mouth or by intravenous drip. A typical life-saving formula is one pinch of salt and one fist of sugar, made up to one litre with clean water.

Prevention
- Supplies of clean drinking water.
- Good hygiene and sanitation.
- Vaccination – although this is only partially successful: immunity is short-lived and not guaranteed. However, booster vaccination rapidly restores immunity.

Carriers
One of the problems with cholera is that some people can be infected without having the symptoms. These carriers of the disease can be very dangerous. Because they think they do not have the disease, they are less likely to take care with sanitary arrangements.

pinch of salt

fist of sugar

1 litre of clean water

Quick test

1 What does HIV stand for?

2 Why should AIDS not be called a disease?

3 What cell does the HIV virus attack?

4 How does it attack this cell?

5 If the HIV virus is so easily destroyed outside the body, why is the disease spreading?

6 Why is the HIV virus so successful?

7 What type of microorganism is cholera?

8 What does it do to gut epithelial cells?

9 What simple medicinal formula can be given to cholera sufferers?

1. human immunodeficiency virus   2. because it is a condition where the immune system is weakened, making the person less able to fight off infection   3. T-lymphocyte   4. It injects RNA and reverse transcriptase enzyme which makes DNA from viral RNA. DNA enters nucleus of T-lymphocyte where the DNA then makes copies of the virus.   5. lack of precautions such as safe sex and discarding used needles   6. It remains dormant for years, thus allowing the unsuspecting victim to infect others.   7. a vibrio bacterium   8. It causes the cells to produce the enzyme adenyl cyclase, which makes the cells lose fluids.   9. a pinch of salt and a fist of sugar in a litre of clean water

Malaria

Malaria is caused by the <u>protozoan parasite</u> *Plasmodium*. **The main symptoms are sweating and fever which occur in bouts as the parasites are released into the blood.**

<u>parasite</u>

<u>red blood cell</u>

Spread

Malaria is spread by a vector, the female *Anopheles* mosquito. The mosquito feeds parasitically on the blood of mammals. When the mosquito has a meal it sucks up some of the **<u>Plasmodium</u>** parasites with the blood. These parasites are then injected into the blood of its next victim.

Cause

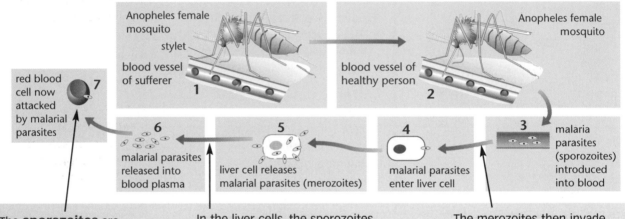

Infection by malarial parasite

Anopheles female mosquito
stylet
blood vessel of sufferer
1

Anopheles female mosquito
blood vessel of healthy person
2

3 — malaria parasites (sporozoites) introduced into blood

4 — malarial parasites enter liver cell

5 — liver cell releases malarial parasites (merozoites)

6 — malarial parasites released into blood plasma

red blood cell now attacked by malarial parasites — 7

The **<u>sporozoites</u>** are carried by the blood to the liver. Here they invade liver cells and disappear completely from the blood.

In the liver cells, the sporozoites multiply to produce **<u>merozoites</u>**. The merozoites are then released back into the bloodstream. This causes the onset of fever.

The merozoites then invade red blood cells. Here they multiply further, releasing merozoites into the blood at regular intervals.

Treatment and prevention

<u>Quinine</u> was the drug originally used to treat malaria. It is obtained from the bark of the cinchona tree and tastes very bitter. British settlers to India used to add gin to their bitter 'tonic' to improve the taste of the medicine. This is where gin and tonic was first drunk. Modern drugs have now replaced the use of quinine but people still drink gin and tonic!

Prevention

With malaria, prevention is better than cure.

- **<u>Use of prophylactic drugs</u>**. People who are going on holiday to areas of the world where malaria is endemic are encouraged to start taking a course of drugs before they go on holiday. If they are subsequently bitten by a malaria-carrying mosquito, the drugs already in the bloodstream will kill the sporozoites.

- **<u>Use of insecticides to kill the mosquitoes</u>**. DDT used to be used as a spray to kill the mosquitoes. Unfortunately the mosquitoes became resistant to the DDT and it was found that the pesticide was being passed into the human food chain. Many countries have now banned its use and other insecticides are now being used.

- **<u>Draining swamps</u>**. The *Anopheles* mosquito lays its larvae in water. Draining the swamps removes one of the main breeding places for the mosquito. This reduces the number of mosquito vectors.

- **<u>Putting fish into swamps</u>**. Sometimes it is not possible to drain the swamps. An alternative is to add fish, such as guppy, to the water. The fish eat the mosquito larvae. This is an example of **<u>biological control</u>**.

Defence against disease

The human body has lots of ways of defending itself.

EXAMINER'S TOP TIP
The following examples show how the body uses external defences against disease. Candidates often forget these when answering a question and just concentrate on internal defences.

The body's defence system

Ears contain wax to stop bacteria infecting the outer ear.

Tears contain the **enzyme lysozyme** that destroys any bacteria that might gain entry to our bodies through our eyes.

The **tough outer layer of skin** is made from the protein **keratin**. The outer layer is dead and provides a barrier against the entry of pathogenic microbes.

Lungs are lined with **cilia** and mucus to trap and remove bacteria.

The **skin** is populated with **'harmless bacteria'** that prevent pathogenic bacteria multiplying too quickly on our skin.

When the **skin is punctured**, the **blood rapidly clots** to seal the wound and **prevent the entry of pathogens**.

Food poisoning
Food poisoning occurs when bacteria are taken in with our food. Some of the bacteria survive the acid in the stomach and start to reproduce in the small intestine and colon. The symptoms of food poisoning can be caused by the chemical toxins the bacteria produce.

The **stomach** contains **hydrochloric acid**. This has a pH of about 3 and is strong enough to kill most **pathogenic bacteria**, most of the time. If it is overwhelmed and fails we can get food poisoning.

EXAMINER'S TOP TIP
Questions often ask you to relate specific ways in which the body can defend itself against specific diseases.

Quick test

1 What is the biological vector for the disease of malaria?

2 How does this vector spread the disease?

3 Which drug was originally used to treat malaria?

4 Why does a person suffering with malaria have regular bouts of fever?

5 How can the spread of malaria be prevented?

6 What are the main external mechanisms that the body uses to protect us from disease?

7 What does the word pathogenic mean?

8 How do bacteria cause the symptoms of food poisoning?

1. Female *anopheles* mosquito 2. The mosquito bites an infected animal and sucks up blood containing the malaria merozoites. When it bites a human, merozoites pass into the human bloodstream. 3. quinine from the cinchona tree 4. The merozoites enter the liver and disappear from the blood. They reproduce then suddenly re-enter the bloodstream, causing the onset of fever. 5. prophylactic drugs to kill all the parasite when it enters the blood of the host; use of insecticides; draining swamps; putting fish into the swamps to eat the mosquito larvae 6. layer of skin to prevent entry of bacteria; blood clots to seal wounds; tears contain the enzyme lysozyme to kill bacteria; mucus and ear wax to trap bacteria; cilia line trachea to move mucus and trapped bacteria up and out of the lungs; stomach acid to kill bacteria to reduce the colonisation by pathogenic bacteria 7. disease causing 8. by producing toxins

Immunity

White blood cells (<u>leucocytes</u>) are the body's main defence against disease. They are made by dividing stem cells in the bone marrow. There are four different types.

Four types of leucocytes

1 <u>Neutrophils</u> (phagocyte)
- ingest bacteria by **<u>phagocytosis</u>**
- live only a few days
- are the first to arrive at an infection site.

2 <u>Macrophages</u> (phagocyte)
- ingest bacteria by phagocytosis
- are longer lived
- take over from neutrophils at an infection site.

<u>Natural immunity</u> is the body's first line of defence.
- It is **<u>non-specific</u>**. The phagocytes will engulf any foreign organism in the body.
- It does not adapt to the type of organism that is invading our body.
- It does not acquire a 'memory' of how to attack the invading organism.

<u>Adaptive immunity</u> is the body's second line of defence.
- It is **<u>specific</u>** against each type of pathogen.
- It is acquired only when we come into contact with the pathogen.
- It has a 'memory' so that the next time we come into contact with the pathogen, our defence mechanism can react very quickly.

3 <u>B-lymphocytes</u>
- are formed in the bone marrow – 'B' stands for 'bone'
- produce **<u>antibodies</u>**.

4 <u>T-lymphocyte</u>
- mature in the thymus gland – T stands for 'thymus'
- do not produce antibodies but help macrophages do their job.

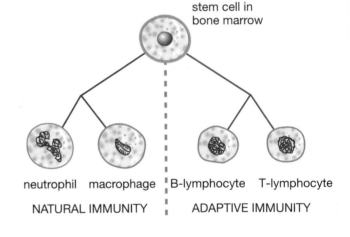

stem cell in bone marrow

neutrophil   macrophage   B-lymphocyte   T-lymphocyte

NATURAL IMMUNITY   ADAPTIVE IMMUNITY

Phagocytes

Phagocytes move through blood, tissue fluid and the lymphatic system. They engulf bacteria. **Pseudopodia** flow around the bacteria forming an internal vacuole. They then digest them using the enzyme lysozyme.

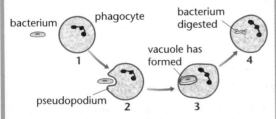

bacterium   phagocyte   bacterium digested

vacuole has formed

pseudopodium

1   2   3   4

Phagocyte engulfing a bacterium

Opsonins

- Opsonins are chemical antibodies that help phagocytes do their job.
- They can attach themselves to microorganisms and attract phagocytes to engulf them.
- They can also stimulate phagocytes to become more active.
- Some can even damage the cell membranes of bacteria.

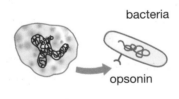

bacteria

opsonin

Opsonin on the surface of a bacterium

EXAMINER'S TOP TIP

This section on immunity can be quite confusing because textbooks often use different words to describe the same thing. Opsonins are sometimes called complement proteins, for example. You would be advised to use one text and stick to it.

Lymphocytes

B-lymphocytes and antibodies

Any substance that initiates an immune response is called an **antigen**.
An **antibody** is a specialised protein molecule. It is specific and will bind with a particular antigen.

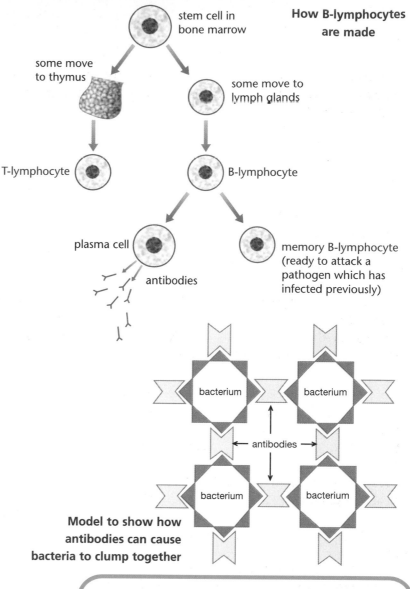

How B-lymphocytes are made

stem cell in bone marrow

some move to thymus

some move to lymph glands

T-lymphocyte

B-lymphocyte

plasma cell

antibodies

memory B-lymphocyte (ready to attack a pathogen which has infected previously)

bacterium
bacterium
antibodies
bacterium
bacterium

Model to show how antibodies can cause bacteria to clump together

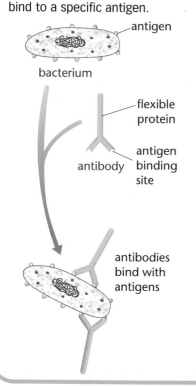

Antibodies are made by one type of B-lymphocyte. An antibody will bind to a specific antigen.

antigen

bacterium

flexible protein

antibody

antigen binding site

antibodies bind with antigens

Antibodies work by:

- **rupturing bacterial cell membranes** (**lysis**) – kills the bacteria and the remains will be engulfed by phagocytes.
- **clumping** – binds bacteria together so that the phagocytes can engulf many in one go.
- **neutralising toxins** – neutralising the chemicals produced by bacteria that make us feel ill and damage cells and tissues in our body.

T-lymphocytes

T-lymphocytes exist as three different types.

- **Helper T-cells** – these help in the production of antibodies and help the B-lymphocytes to work. The HIV virus attacks these cells.
- **Killer T-cells** – these destroy human cells that have become infected with a virus.
- **Suppressor T-cells** – these suppress other cells in the immune system. They switch the defence system off. This is to prevent it over-reacting after the antigen has been destroyed.

Quick test

1 What is natural immunity?

2 What is adaptive immunity?

3 What are opsonins?

4 How do phagocytes engulf bacteria?

5 What do antibodies do?

6 Name three types of T-lymphocytes.

7 What does the letter T in T-lymphocytes mean?

8 Why are killer T-cells so named?

Use the questions to test your progress. Check your answers on pages 94–95.

Human health and disease

1 a Describe five different causes of disease. [5]

....

....

b Describe how the disease of tuberculosis can be prevented. [4]

....

....

2 AIDS is caused by the HIV virus.
a Put the following pictures in their correct order. ☐ ☐ ☐ [2]

i   ii   iii

b Explain why HIV is such a successful virus even though it kills the host in which it lives. [1]

....

c Explain why it is difficult for the body's immune system to fight the HIV virus. [1]

....

d Describe three ways that HIV can be passed from one person to another. [3]

....

....

3 Malaria is a disease that affects millions of people around the world.
a Explain why malaria cannot be passed from one person to another by direct contact. [1]

....

b Look at the picture of blood as seen under a microscope.
Say whether the person who donated this blood has
malaria and explain why. [1]

....

....

....

....

c Describe four ways in which the spread of the malarial parasite can be prevented. [4]

....

....

....

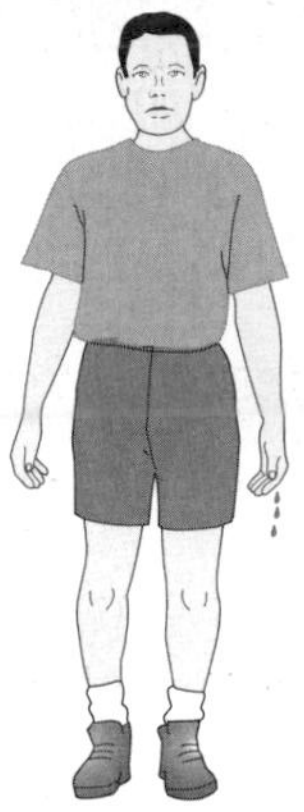

4 Copy the picture of the human body and label it to show the ways that
the human body protects itself from disease. [5]

5 The diagram shows how different cells are produced from stem cells
in the bone marrow.
a Explain the difference in terms of immunity between the two cells on
the left and those on the right. [2]

....

....

b Explain the difference between the two types of
immunity provided by the cells. [3]

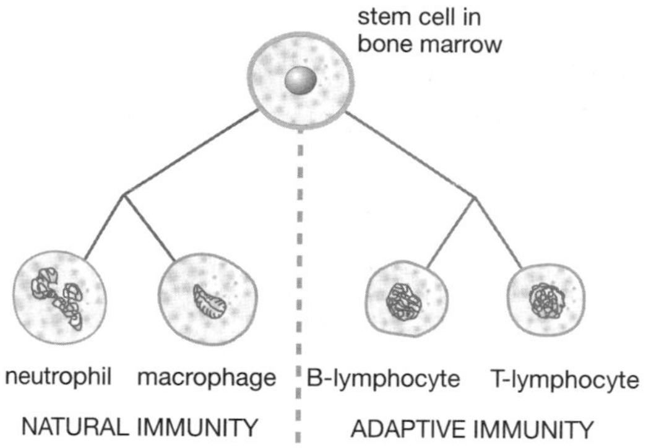

....

....

....

stem cell in
bone marrow

6 The diagram shows how the immune system responds
to a foreign antigen.
a Describe the role of the B-lymphocyte in the immune
response. [2]

neutrophil   macrophage   B-lymphocyte   T-lymphocyte

NATURAL IMMUNITY   ADAPTIVE IMMUNITY

....

....

b Some B-lymphocytes turn into memory cells.
Explain the advantage of this to the human body. [2]

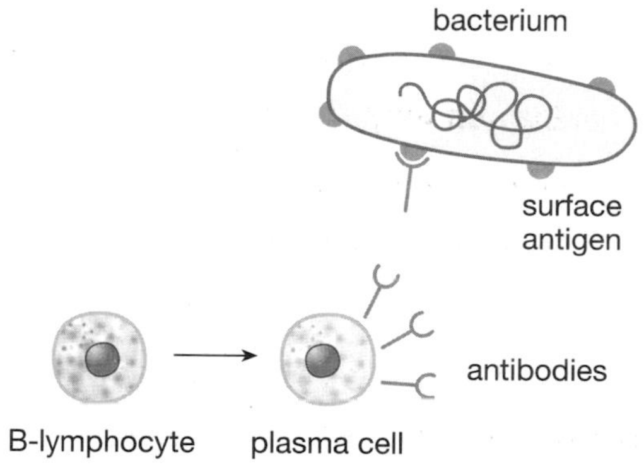

bacterium

....

....

surface
antigen

c People can catch the common cold again
and again. Explain why some diseases are only
caught once, while others, such as the common cold
virus, can be caught many times. [2]

antibodies

....

B-lymphocyte   plasma cell

....

d Explain the role of the T-lymphocyte in the immune system. [3]

....

....

e Explain why antibodies that bind several bacteria together are useful to the
immune system. [2]

....

....

f Describe one other role of antibodies. [1]

....

....

....

Total: /44

Ecosystems

The biosphere and biomes

<u>Ecology</u> is the study of the relationships between living organisms and their environment.

The part of the Earth that supports life is a thin layer called the <u>biosphere</u>.

The biosphere contains areas such as grassland, forests or seas. These areas are called <u>biomes</u>.
Within the biome are <u>ecosystems</u>.
Within the ecosystem are <u>communities</u>.
Communities contain <u>populations</u> which consist of individual organisms belonging to a single species.

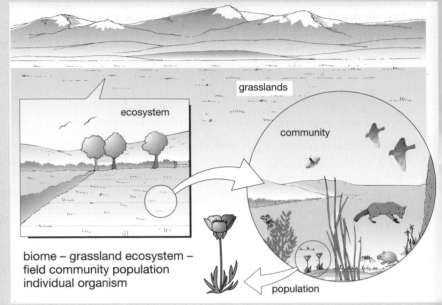

biome – grassland ecosystem – field community population individual organism

Ecology is usually studied by looking at specific <u>ecosystems</u>. An ecosystem is a stable ecological unit – for example, a lake, a field or a wood. An ecosystem consists of two parts.

Factors influencing the ecosystem

Biotic factors – living
These include all the inter-relations between the organisms in the environment. They include:
- **predation** – one organism preying upon another
- **parasitism** – one organism living on or inside another organism, at the expense of that organism and giving nothing in return
- **commensalism** – one organism living off another organism that does not lose or benefit from the relationship
- **symbiosis** – two organisms living together for mutual benefit
- **competition** – any number of organisms in competition for resources.

These two groups of factors all influence how the ecosystem functions.

Abiotic factors – non-living
These are non-living factors that affect organisms in the environment. They include:
- **water** – needed by all living things
- **temperature** – most organisms can survive between 0 °C and 40 °C, but some can survive outside this range
- **light** – needed for photosynthesis and supplies energy for other organisms in the food web
- **atmosphere** – includes wind and concentrations of oxygen and carbon dioxide
- **edaphic factors** – soil conditions such as pH, soil type and drainage.

EXAMINER'S TOP TIP
Don't think of ecosystems as self-contained units. They merge one into the other without fixed physical boundaries.

Energy and ecosystems

Ecosystems

Most ecosystems obtain their energy from **photosynthesis**. Photosynthesis involves using the energy from sunlight to make complex organic molecules such as carbohydrates. These in turn can be converted into lipids and proteins.

Energy from light causes **chlorophyll** to release an excited electron.

Calvin cycle – this is a cyclic reaction. ATP and reduced **NADP** (NADPH$_2$) along with carbon dioxide from the atmosphere enter the cycle.

Oxygen released into atmosphere

light-dependent stage (in the thylakoid membranes)

light-independent stage (in the stroma)

light

$4OH^- = 2H_2O + O_2 + 4e^-$

chlorophyll

OH^-

A 6-carbon sugar such as glucose is produced.

$H_2O =$

H^+

6C sugar

ATP

CO_2 from atmosphere

electron carrier system

ATP

triose phosphate

ribulose bisphosphate (RuBP)

e^-

CO_2

Water is split (photolysis) into H^+ and OH^-. Electrons are released when $4OH^-$ is converted into water and oxygen. The electrons released replace those lost by the chlorophyll.

e^-

glycerate-3-phosphate (GP)

NADPH$_2$

NADP is reduced by the electron and a hydrogen ion produced by the photolysis of water.

NADPH$_2$ becomes oxidised and is then reused in the light stage.

As the electron is passed along an **electron carrier system** it releases energy, producing **ATP**.

Not all of the energy in the sunlight that lands on green plants is used by photosynthesis. Some of it is used to heat up the plant. Some of it is reflected straight back off. Some of it misses the chloroplasts. Only about 5% of the light energy gets used in photosynthesis.

Quick test

1. What is the name of the thin layer that covers the whole of the Earth's surface that supports life?
2. What are similar areas called within this thin layer?
3. What is an ecosystem?
4. List five biotic factors.
5. List five abiotic factors.
6. What is meant by edaphic factors?
7. Where does the light-dependent stage of photosynthesis take place?
8. Where does the light-independent stage of photosynthesis take place?
9. What three things are used by the Calvin cycle?
10. Where does the energy to produce the ATP come from in the light-dependent reaction?

1. biosphere 2. biomes 3. a stable ecological unit within a biome, such as a pond or a wood 4. predation, parasitism, commensalism, symbiosis, competition 5. water, temperature, light, atmosphere, edaphic 6. factors to do with the soil, such as pH, composition and drainage 7. thylakoid membranes 8. stroma 9. carbon dioxide, ATP, reduced NADP (NADPH$_2$) 10. Light excites an electron which is released from a chlorophyll molecule. The energy in the excited electron is used to produce ATP from ADP.

Energy flow through ecosystems

The **flow of energy** through an ecosystem can be shown in **several different ways**. You will have studied all of these for your GCSE. However, you need to know how they relate to each other and understand that each one is showing the flow of energy in a different way.

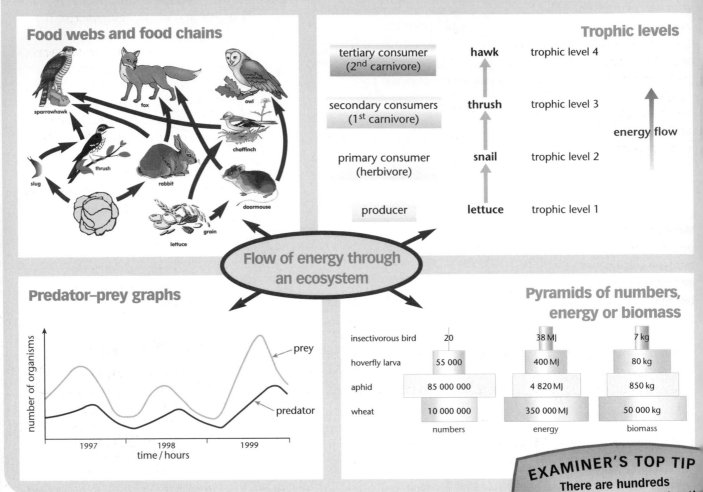

Food webs and food chains

sparrowhawk
fox
owl
chaffinch
thrush
slug
rabbit
doormouse
grain
lettuce

Flow of energy through an ecosystem

Trophic levels

tertiary consumer (2nd carnivore)	**hawk**	trophic level 4
secondary consumers (1st carnivore)	**thrush**	trophic level 3
primary consumer (herbivore)	**snail**	trophic level 2
producer	**lettuce**	trophic level 1

energy flow

Predator–prey graphs

number of organisms

prey

predator

1997   1998   1999
time / hours

Pyramids of numbers, energy or biomass

	numbers	energy	biomass
insectivorous bird	20	38 MJ	7 kg
hoverfly larva	55 000	400 MJ	80 kg
aphid	85 000 000	4 820 MJ	850 kg
wheat	10 000 000	350 000 MJ	50 000 kg

Cycles within ecosystems 1

Ecosystems are self-sustaining. One of the reasons is because they recycle nutrients. There are two examples that you need to know.

The carbon cycle

The **carbon cycle** can also be used to show the **flow of energy**. The arrows show how energy is passed on from one organism to another. Energy is at its lowest when **carbon dioxide** is in the air and the whole cycle starts when energy from **sunlight** is absorbed for **photosynthesis**.

Combustion of fossil fuels is releasing increasing quantities of carbon dioxide leading to global warming.

The atmosphere contains about 0.04% carbon dioxide. This is slowly rising due to the burning of fossil fuels.

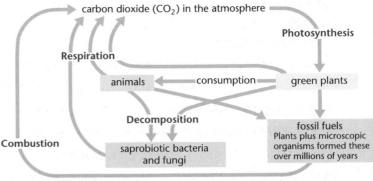

carbon dioxide (CO_2) in the atmosphere

Photosynthesis

Respiration

animals ← consumption ← green plants

Decomposition

Combustion

saprobiotic bacteria and fungi

fossil fuels
Plants plus microscopic organisms formed these over millions of years

Saprophytes release carbon dioxide by decomposition of dead organisms.

Millions of tonnes of carbon are also locked up in carbonates in rocks. These rocks were formed from the bodies of microscopic sea creatures.

Cycles within ecosystems 2

The nitrogen cycle

Nitrogen is used to make proteins. Unfortunately it is a very unreactive gas. and animals and plants cannot use it directly from the air, even though the atmosphere contains 79% nitrogen.

Bacteria such as **Pseudomonas** and **Azobacter** convert some of the nitrates back into atmospheric oxygen. They are anaerobic bacteria and live in soils short in oxygen.

Only certain types of bacteria can 'fix' nitrogen from the air and make nitrogen-containing compounds. **Rhizobium** bacteria have a **symbiotic** relationship with plants. They make nitrates and in return get carbohydrates from the plants.

Nitrification is the process where nitrogen is converted into nitrates that can be used by plants.

nitrogen gas (N_2) in the atmosphere

Denitrification (*Pseudomonas* bacteria)   lightning   **Nitrogen fixation**

free living nitrogen fixing bacteria in soil (*Azotobacter*)

other green plants

animal protein ← eaten by animals

death   faeces   urine

Decomposition by saprobiotic bacteria and fungi

In legumes (pea and bean family) *Rhizobium* **bacteria** live in nodules in the roots and produce NH_4^+ **ions** which enables the plant to make amino acids then **proteins**. Some carbohydrates are used by the bacteria.

artificial fertiliser

NO_3 (nitrate) ← *Nitrobacter* bacteria ← NO_2 (nitrite) ← *Nitrosomonas* bacteria ← NH_3 (ammonia)

← **Nitrification** →

Lightning also converts nitrogen in the air into nitrates in the soil.

It is possible that genetic engineering could be used to grow plants that were able to fix nitrogen, just like **Rhizobium** bacteria. Farmers would save a fortune in not having to use fertilisers containing nitrates and food would be much cheaper.

Some very important social decisions involving complex issues are now being taken about the future use of genetically modified crops.

EXAMINER'S TOP TIP
There are hundreds of different ways of drawing the nitrogen cycle. Remember just one, or better still understand the process so that you can draw your own.

Quick test

1 *Name three different ways that you can show the flow of energy through an ecosystem.*
2 *What happens to the amount of available energy as it passes through an ecosystem?*
3 *Carbonate rocks have been built up over millions of years from the bodies of small sea creatures. Why have the levels of carbon dioxide in the atmosphere been dropping over the same time period?*
4 *Name two chemical processes that return carbon dioxide to the atmosphere.*
5 *Why could massive deforestation lead to increased levels of carbon dioxide in the atmosphere?*
6 *How do saprophytes contribute to increased levels of carbon dioxide?*
7 *Why do animals and plants not use nitrogen directly from the air?*
8 *How do plants obtain a source of nitrogen?*
9 *What effect of the weather can convert nitrogen gas into nitrates?*

1. food webs and food chains; pyramids of number and biomass; trophic level charts and tables 2. It is reduced as energy is lost at each trophic level. 3. Carbon has been trapped in the carbonate rocks, so less is available to form carbon dioxide. 4. burning, respiration 5. Trees use carbon dioxide from the atmosphere for photosynthesis. 6. They decompose dead organisms to release carbon dioxide through respiration. 7. Nitrogen is very unreactive. 8. from fertilisers; decay of dead organisms; nitrogen-fixing bacteria in roots and soil 9. lightning

Increasing yield in agriculture

Farmers use various agricultural methods to produce food as efficiently as possible.

Fertilisers

Fertilisers are a quick but expensive way of providing plants with the nutrients that they need.
Fertilisers usually contain NPK –
nitrogen, phosphates and potassium.

- **Nitrogen** is required for making proteins and growth. Plants lacking nitrogen are stunted.
- **Phosphates** are used to produce ATP and DNA. Plants lacking phosphates have purplish leaves.
- **Potassium** is used to help make proteins and chlorophyll. Plants lacking potassium have their older leaves turnng yellow.

Greenhouses

Greenhouses enable farmers to have more control over the environment. They can provide better growing conditions by controlling

- **temperature** – can be kept at optimum for the crop
- **water and nutrients** – computer-controlled irrigation provides optimum levels of water; the nutrients supplied can match the needs of the crop
- **carbon dioxide** – the enclosed area allows carbon dioxide levels to be increased; as carbon dioxide is a limiting factor, this will increase the rate of photosynthesis and plant productivity.

Increase yield

> **EXAMINER'S TOP TIP**
> Don't try to remember all these methods individually.
> Learn them in groups.

Pest control

There are many methods that farmers can use to control pests.

- **Pesticides** – may kill pests by direct contact but some are **systemic**. Systemic pesticides are absorbed by the plants and kill the pest when it feeds on them.
- **Biological control 1** – releasing predators that destroy the pest. This is often best done in greenhouses where predators can be contained. Whitefly is a major pest in greenhouses; **Encarsia** wasps lay their eggs in whitefly larvae; the eggs hatch into larvae which feed on the whitefly larvae.
- **Biological control 2** – sex pheromones can be used to attract insect pests, which can then be destroyed.
- **Biological control 3** – releasing sterile males of the pest. Male pests are sterilised by being irradiated with a radioactive source. Females try to breed with them but are unsuccessful, thus reducing the numbers of the pest.
- **Biological control 4** – genetic engineering. Crops can be genetically engineered so that they contain a gene that produces toxins that kill the pests. However, the toxins must not be dangerous to humans or spoil the quality of the crop.

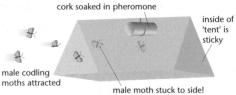

cork soaked in pheromone

inside of 'tent' is sticky

male codling moths attracted

male moth stuck to side!

A codling moth trap

Herbicides and weed control

The crop plants require water, light and minerals. Weeds compete for these resources. The more resources the weeds get, the less is available for the crop. Weeds can be controlled by various methods.

- **Herbicides** – sometimes called weed killers. Unfortunately, they also kill the crop plants, so herbicides are often used before the crop is planted. They are called **pre-emergent** weed killers.
- **Genetic engineering** has now produced some crop plants that are resistant to the herbicide. This means that the crop can be sprayed with the herbicide while it is growing. The weeds are killed but not the crop.
- **Mechanical control** – this includes everything from pulling up weeds by hand to hoeing or using rotorvators and specialised machinery.

Succession of set-aside land

Farmers are now being encouraged to set surplus land aside and allow it to develop naturally. This land is called 'set aside'.

Stage 1 – Primary colonisers

These are plants that can quickly exploit a new habitat. They are usually annual plants such as grass and weeds. The seeds are brought by wind and birds and the plants grow to produce new seed that can quickly germinate in spring of the following year. When the original plants die in the autumn, they decay to provide nutrients for next year's growth.

Stage 2 – Secondary colonisers

These are plants that take more time to become established. Many are perennial and do not die off completely in the winter. This gives them a head start in the spring and eventually they start to replace some of the primary colonisers. They also include plants such as small shrubs and bushes that begin to appear when the land has been left for a couple of years.

Stage 4 – Climax community

After many years, larger trees such as oak become established. They form a dense canopy as they compete for light. This produces a very stable climax community, which can exist in this state for hundreds of years. The community changes as new plants and animals colonise the new habitat, forming very complex but stable food webs.

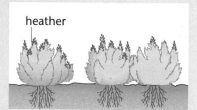

Stage 3 – Birch woodland

After several years the shrubs begin to be replaced by trees such as birch. Birch has a fast-growing habit and can reach the light above the shrub layer. It does not have a dense canopy but some light to the lower layers is restricted and they gradually reduce in density. Very early flowering plants, such as bluebells, begin to colonise the ground layer as they can grow and flower before the tree layer comes into leaf.

Different ecosystems will end up with different climax communities. The one described above is just an example.

Quick test

1 **What three minerals are usually found in fertiliser?**
2 **List four methods of biological pest control.**
3 **How can greenhouses be used to increase crop yield?**
4 **Why does increasing levels of carbon dioxide result in a greater crop yield?**
5 **Why have some crop plants been genetically engineered to be resistant to weed killers?**
6 **What is meant by 'set-aside land'?**
7 **What are the primary colonisers of bare earth?**
8 **What is the final community called?**
9 **What is the process called that starts with primary colonisers and ends up with a climax community?**

1. nitrates, phosphates, potassium 2. releasing predators to the pest; using pheromones to attract the pest; releasing sterile males; genetically engineering crops to produce natural pesticides 3. by controlling environmental factors such as water, nutrients, temperature and levels of carbon dioxide 4. Carbon dioxide is a limiting factor for photosynthesis so an increase in levels increases the rate at which photosynthesis occurs. 5. so the weed killer can be used to kill weeds without harming the crop 6. land that farmers do not use but allow to revert to a natural state, encouraging wildlife to return 7. mosses and annual weeds 8. climax community 9. succession

Eutrophication

Human activity has had a significant effect on the environment. It has destroyed habitats and caused various kinds of pollution. <u>Eutrophication</u> is one type of water pollution.

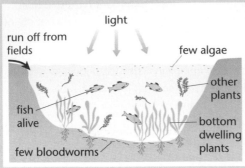

light

run off from fields

few algae

other plants

fish alive

bottom dwelling plants

few bloodworms

Cross section of river before eutrophication

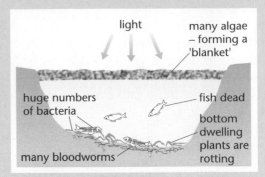

light

many algae – forming a 'blanket'

huge numbers of bacteria

fish dead

bottom dwelling plants are rotting

many bloodworms

Cross section of river after eutrophication

EUTROPHICATION

The river has:
- a large number of organisms
- a wide range of different species
- a high level of oxygen in the water
- low biological oxygen demand (BOD).

Addition to the river of:
- sewage
- fertiliser from fields
- nitrates and phosphates from washing powders.

The river has:
- a reduced number of organisms
- a smaller range of species
- a low level of oxygen in the water
- high biological oxygen demand (BOD).

EXAMINER'S TOP TIP
Eutrophication is all about how much oxygen is in the water

How it happens

Stage 1
Pollutant enters the river. It contains minerals and **nutrients** that encourage plant growth. The effect is just like adding a fertiliser to the water.

Stage 2
The nutrients encourage the growth of **green algae**. The algae floats near the surface and soon forms a thick green mat covering the surface of the pond or slow-moving river.

Stage 3
The thick mat of green algae blocks out the light from the sun to the water below. This means that water plants cannot photosynthesise and they die.

Stage 6
Eventually the only animals that can live in the water are **bloodworms** (**Tubifex**). These contain a red pigment similar to haemoglobin.
It allows them to get just enough oxygen from the stinking mud at the bottom of the pond to survive. Their numbers also increase because they are no longer being eaten by the fish.

Stage 5
Fish die as there is no longer sufficient oxygen for them to breathe. As they begin to rot, the bacteria use up the remaining oxygen. Soon the river is full of rotting animals and plants.

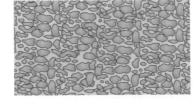

Stage 4
Bacteria start to decay the dead plants. The rotting plants provide food for the bacteria and they rapidly increase in number. The dead plants are no longer producing oxygen. What oxygen is left in the water, is soon used up by the bacteria.

Over-fishing

Fish are one of our major sources of food. They are rich in protein and unsaturated fats. But the increase in fishing and modern fishing techniques decreased fish stocks.

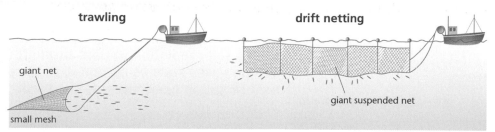

trawling — giant net — small mesh

drift netting — giant suspended net

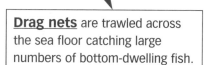

Drift nets hang down into the sea like curtains, trapping large numbers of surface-dwelling fish.

Drag nets are trawled across the sea floor catching large numbers of bottom-dwelling fish.

Small mesh nets have a mesh so small that immature fish are caught. This reduces the fish stock even further as these fish have not been allowed to spawn.

How to conserve fish stocks

Legislation

Legislation has been introduced to provide quotas to different fishermen in different countries. Fishermen are prevented, by law, from catching more fish than their quota allows. Exclusion zones may be introduced to prevent any fishing and to allow stocks to recover. Laws on mesh sizes in nets may prohibit the use of nets with a small mesh size.

International agreements

Different countries can agree to stop fishing in certain areas to allow the stocks to return to their normal numbers, but such agreement can be hard to achieve.

Fish farms

These can be used to breed fish in large numbers. However, so many fish in close proximity to each other can lead to an increase in disease and parasites.

Quick test

1 List two pollutants that can cause eutrophication of rivers.
2 What does BOD stand for?
3 Why do fish die after eutrophication?
4 Explain why eutrophication leads to a lack of oxygen.
5 Name two types of net used to catch fish.
6 List three ways that can be used to increase fish stocks.
7 What is one problem that can occur when keeping fish in large numbers on a fish farm?

1. fertilisers; washing powders; sewage; nitrates 2. biological oxygen demand 3. lack of oxygen 4. Fertilisers cause algae to grow; algae blocks out sunlight; plants die and rot; bacteria feed on rotting plants and use up all the oxygen. 5. drift nets and drag nets. 6. introduce legislation to give quotas to stop over-fishing; international agreements, e.g. fishing-free zones; fish farming. 7. increase in disease / parasites

Human activity and the atmosphere

The greenhouse effect

Even on a cold day, a greenhouse is always warm if the sun is shining. This is called the 'greenhouse effect'.

It happens because:

- Glass is transparent to solar radiation, allowing it to pass through.
- The solar radiation causes the surfaces in the greenhouse to warm up.
- The surfaces emit heat radiation.
- Glass is not transparent to heat radiation so it is reflected back into the greenhouse, warming it up.

A similar thing happens with the Earth's atmosphere. Just like the glass, the atmosphere allows solar radiation to pass through. However, certain gases such as carbon dioxide reflect the heat radiation back towards the Earth.

Greenhouse gases

- **Water vapour** – levels increase as the temperature of the Earth increases, causing a spiral effect.
- **Carbon dioxide** – levels are increasing as we burn more fossil fuels.
- **Methane** – produced from rotting matter and flatulent cattle.

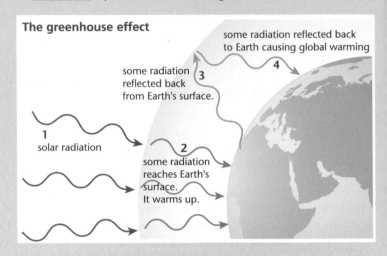

The greenhouse effect

The effects of this **global warming** include changing weather patterns and a rise in sea levels as the polar icecaps start to melt.

Heat in a greenhouse

EXAMINER'S TOP TIP
Global warming is all about the balance between how much energy the earth is receiving and how much it is losing.

The hole in the ozone layer

- The **ozone layer** is a thin layer of ozone gas that surrounds the Earth.
- The sun gives out large amounts of **ultraviolet radiation**. This radiation is extremely dangerous to living organisms. It can damage DNA and cause skin cancer.
- The ozone layer filters out large amounts of this ultraviolet radiation. This is why life can exist on the Earth's surface.
- Some chemicals, such as **chlorofluorocarbons** (CFCs) that are used in aerosols and fridges, react with the ozone.
- This has caused holes to appear in the ozone layer.
- So far the holes have been over the Earth's poles where there is little life.
- It is important to reduce the use of chemicals that destroy the ozone layer so that the ozone layer can reform and provide protection for life on Earth.

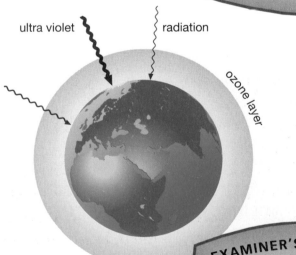

EXAMINER'S TOP TIP
Many candidates confuse global warming with the hole in the ozone layer and lose marks. The fact that some gases can be both greenhouse gases and ozone layer destroyers does not help!

Acid rain

Acid rain is caused when fossil fuels, containing sulphur, are burnt.

Fossil fuels contain small amounts of **sulphur**.

$$S + O_2 \longrightarrow SO_2$$

When the fossil fuels are burnt, **sulphur dioxide** gas is formed.

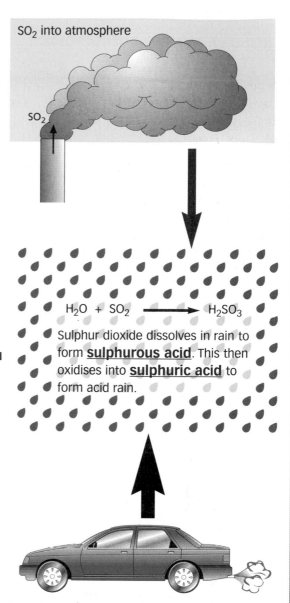

SO_2 into atmosphere

SO_2

$$H_2O + SO_2 \longrightarrow H_2SO_3$$

Sulphur dioxide dissolves in rain to form **sulphurous acid**. This then oxidises into **sulphuric acid** to form acid rain.

Car exhaust fumes contain gases such as oxides of nitrogen (NOX). These dissolve in rain to form **nitric acid** rain.

<a/w 361>

Acid rain starts to kill trees and damage limestone buildings.
Acid rain also lowers soil pH. This reduces the solubility of many minerals, making them unavailable to plants.

Lichens can be used as acid rain indicators. The number and types of species present can be used to determine the level of pollution caused by acid rain.

Lots of different species of lichens $\longrightarrow$ increasing acid rain pollution $\longrightarrow$ Few species of lichens

Quick test

1 **Name three greenhouse gases.**
2 **State two effects of global warming.**
3 **How do greenhouse gases cause global warming?**
4 **How does the ozone layer protect the Earth?**
5 **Name one kind of chemical that can destroy the ozone layer.**
6 **Name two sources of the gases that cause acid rain.**
7 **Write down the chemical formula for the burning of sulphur to produce sulphurous acid.**
8 **What biological indicator can be used to determine the level of acid rain pollution?**

1. water vapour, carbon dioxide, methane 2. changing weather patterns and rising sea levels 3. Solar radiation enters the atmosphere and warms up the Earth's surface; some solar radiation is reflected back into space; greenhouse gases prevent this reflected radiation from escaping back into space. 4. It stops large amounts of ultraviolet radiation from reaching the Earth's surface. 5. CFCs found in aerosols and as coolants in fridges 6. burning fossil fuels (sulphur dioxide) and car exhaust fumes (oxides of nitrogen) 7. $H_2O + SO_2 = H_2SO_3$ 8. number and types of lichen species

Exam-style questions
Use the questions to test your progress. Check your answers on pages 94–95.

Ecology and the environment

1 a List five abiotic factors that affect an ecosystem. [5]

....

b State what is meant by the word ecosystem. [1]

....

c Put the following words in order, starting with the largest. [3]

community   biosphere   ecosystem   biome

2 The diagram to show photosynthesis has labels missing.

a Fill in the missing labels, 1 to 5. [5]

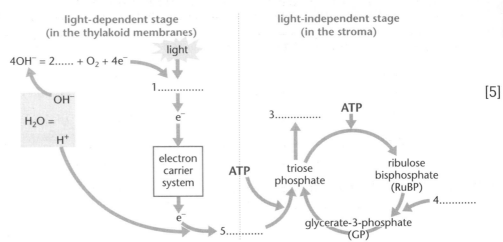

light-dependent stage (in the thylakoid membranes)

light-independent stage (in the stroma)

$4OH^- = 2......+ O_2 + 4e^-$

light

OH^-

$H_2O =$

H^+

1..............

e^-

electron carrier system

e^-

5............

ATP

3..............

ATP

triose phosphate

ribulose bisphosphate (RuBP)

4............

glycerate-3-phosphate (GP)

b State where the light-dependent stage takes place. .... [1]

c State where the light-independent stage takes place. .... [1]

3 The following article appeared in a history textbook.

> **ANCIENT ROMANS DISCOVER NITROGEN CYCLE**
> Farmers in ancient Rome knew that to get the best yield, they should only grow crops in a field for two years. During the third year they would grow a leguminous plant such as clover and then plough it back into the ground. They knew that by doing this, they would get a much better crop the following year. Unfortunately the Romans never did discover why this farming method was so successful.

a Explain why growing the same crop every year in the same field will reduce the yield. [1]

....

b Explain why growing a leguminous plant such as clover and ploughing it in improved the yield the following year. [2]

....

....

c Suggest why modern-day farmers use fertilisers rather than growing clover every third year. [2]

d Explain why nitrates are so important for plant growth. [2]

....

....

e Describe two ways, other than by absorption by plants, that nitrates may be removed from the soil. [2]

....

....

4 Each of the following is a method that can be used for biological pest control.

  i Insect pheromone sex attractants   ii sterile males   iii insect predators

  Describe how each method can be used to reduce insect pests.   [3]

....
....
....

5 The following article appeared in a local newspaper.

> **CROPS MADE RESISTANT TO WEED KILLER**
> A variety of wheat has now been genetically engineered to be resistant to a specific type of weed killer. A gene from a resistant plant has been transferred into the wheat so that the modified wheat is no longer damaged when the field is sprayed with weed killer.

  a Suggest what advantage this modified wheat is to the farmer.   [2]

....
....

  b Suggest why a large agricultural company would want to produce genetically modified wheat. [2]

....
....

6 Ponds, lakes and rivers can be damaged by eutrophication.
  Look at the pictures of before and after eutrophication.

  Explain how fertiliser run-off from a farmer's field can cause so much damage.   [5]

....
....
....

7 The graph shows the size of the population and average length of codfish in the North Sea, over a number of years.

  a Suggest why the average length of the fish is now causing concern to conservationists.   [1]

....

  b Explain why trawling and drift netting are having such an effect on the codfish population. [1]

....

  c Describe three strategies that could be used to help increase the fish stocks to their previous levels.   [3]

....
....
....

Total: /42

Answers

Biological molecules (pp. 4-13)

1 a

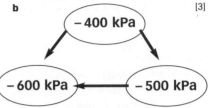

[water produced = 1 mark]
[correct bond = 1 mark]
[correct labels = 1 mark]

b condensation [1]

c glycosidic bond [1]

d

straight chains [1]
cross links between chains [1]

2 a carbon, hydrogen, oxygen [1 mark each]

b proportionately less oxygen [1]

c

[ester bond correct = 1 mark]
[correct labels = 1 mark]

d condensation [1]

e ester bond [1]

3 Saturated fats have no double bonds. [1]
C=C [1]

4 a phosphate group [1]
fatty acid removed from triglyceride [1]
both bond together [1]

b polar head [1]
attracts water [1]
fatty acid tails [1]
repel water [1]

c **phospholipid molecule =**
hydrophobic tail / hydrophilic head

double layer [1]
correct orientation [1]

5 a nitrogen, sulphur, iron, phosphorus
[1 mark each]

b nitrogen [1]

c amino acids; polypeptides [1]

6 a

amine group ... carboxylic acid group

[1 mark each]

b

[peptide bond correct = 1]
[water molecule = 1]

c condensation [1]

d peptide bond [1]

7 a 1 peptide bond; 2 hydrogen bond;
3 amino acid [1 mark each]

b alpha helix [1]

8 a gives strength [1]

b interwoven strands [1]
alpha helix strands [1]

Cells (pp. 16-25)

1 The membrane is fluid. [1]
It looks like a mosaic. [1]

2 a 1 mitochondrion; 2 Golgi body; 3 nucleus;
4 rough endoplasmic reticulum
[1 mark each]

b 1 site of cellular respiration;
2 produces carbohydrates and
glycoproteins;
3 contains DNA;
4 transport and ribosomes are site of
protein production [1 mark each]

3 Prokaryotes have DNA not enclosed in
nucleus; cell wall made of mucopeptides; no
mitochondria or Golgi bodies; no
endoplasmic reticulum. [1 mark each]

4 It provides support against hydrostatic
pressure. [1]
It has plasmodesmata passing through to
enable cytoplasm to connect between cells.

5 a 1 stroma; 2 outer membranes; 3 grana [3]

b grana thyalakoid membranes [1]

6 Cells are prepared by being cooled in a
buffer solution. [1]
The cells and solution are homogenised
in a blender. [1]
The suspension is ultracentrifuged to
separate the organelles. [1]

7 a An electron microscope uses
electrons, not light; is viewed on a screen,
not directly; cannot view living material;
has more powerful magnification.
[1 mark each]

b 1 eyepiece lens; 2 objective lens;
3 stage; 4 condenser lens [1 mark each]

8 They need specialisation of cells. [1]
As cells increase in size, the surface area is
not large enough for the transport of
material into and out of the cell. Volume
increases to power cubed. Surface area
increases to power squared. [1]

9 1 intrinsic; 2 extrinsic [1 mark each]

10 a The channels are permanently open. [1]
The channels are specific for a particular
molecule. [1]

b The hormone opens the channel; [1]
by attaching to receptor site; [1]
The channels are specific for a
particular molecule. [1]

11 It has a glycoprotein on surface, [1]
called an antigen. [1]
It has specific receptor site to recognise
other antigens. [1]

12 Diffusion – random movement of molecules
from high to low concentration [1]
Facilitated diffusion – channel proteins
enable diffusion across plasma membrane
[1]

Active transport – ATP used to provide
energy to move molecules against a
concentration gradient. [1]
Pinocytosis – substances engulfed by plasma
membrane to form a vacuole. [1]

13 a water potential
= solute potential + pressure potential [1]

b [3]

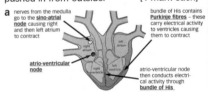

Exchange and transport (pp. 28-39)

1 a thin; moist; large surface area; good
blood/transport supply [1 mark each]

b Any three from: thin; air spaces for
transport of gases; chloroplasts near to
upper surface for light; stoma to allow
exchange of gases during day but reduce
water loss at night; xylem and phloem
for transport. [1 mark each]

2 Con current can only absorb 50% maximum
as the two solutions will then be in
equilibrium. [1]
Counter current maintains a diffusion
gradient so that more than 50% can be
absorbed. [1]

3 The respiratory surface is moist so gas
dissolves. [1]
There is more oxygen in lungs (20%) than in
blood, so oxygen passes along a diffusion
gradient from lungs to blood. [1]

4 The diaphragm contracts and lowers and the
intercostal muscles pull the ribs up and out.
This increases volume of the thoracic cavity.
It also reduces pressure of the thoracic
cavity.
External air pressure is now greater, so air is
pushed in from outside. [1 mark each]

5 a nerves from the medulla
go to the **sino-atrial node** causing right
and then left atrium
to contract

bundle of His contains
Purkinje fibres – these
carry electrical activity
to ventricles causing
them to contract

atrio-ventricular node

atrio-ventricular node
then conducts electri-
cal activity through
bundle of His

[1 mark each]

6 Situated in the right atrium, it helps to
regulate the beating of the heart. [1]

7 a heart muscle relaxing [1]

b heart muscle contracting [1]

8 a 0.6 seconds [1]

b 20 kPa [1]

c Volume decreases as pressure increases
[2]

9 At the arterial end, the blood pressure is
high so tissue fluid leaves the capillary. [1]
At the venous end the blood pressure is
low, so some water returns to blood by
osmosis. [1]

10 1 CO_2 2 H_2CO_3 3 HHb 4 O_2 [1 mark each]

11 Line for foetal haemoglobin starts and
finishes at same place, but rest of line is to
the left of the line for adult haemoglobin. [2]

12 Symplast – water moves from cell to
cell by osmosis. [1]
Apoplast – water moves between cells
because of the cohesion between the
water molecules. [1]

13 Any three from: sunken stomata; hairs to reduce movement of water vapour; curled to retain water vapour; thick cuticle. hairs reduce turbulence. [3]

14 Water moves into source by osmosis. [1]
Water moves up the xylem. [1]
Pressure in source forces sugar solution down phloem to sink. [1]
Sugar converted to starch in sink. [1]
Water potential in sink is high so no water enters. [1]

Reproduction (pp. 42-53)

1 a anaphase [1]
  b iii, iv, ii, i [3]
2 ii, v, i, iv, iii [4]
3 a drawing b [1]
  b It has one chromosome from each pair. [1]
4 a Yellow area are all (n) or haploid. White are all (2n) or diploid. [2]
  b Meiosis takes place between the sporophyte adult and the spores. [1]
5 v, i, ii, iv, vi, vii, viii, iii [7]
6 a No, because the progesterone level has fallen. [1]
  b It would increase the level of oestrogen. [1]
  c about day 14 [1]

Chromosomes, genes and DNA (pp. 56-65)

1 a 5 [1]
  b It has more than one code. [1]
  c GCGCUAGACUCUUAC [1]
  d arginine, aspartic acid, leucine, arginine, methionine [1]
2 a storing information; replication [1 mark each]
  b Conservative hypothesis says that all the new DNA is copied. [1]
  Semi-conservative hypothesis says that only one strand was copied and one was original. [1]
3 a deletion [1]
  b All following triplets will also be affected, because of the displacement of each base. [2]
  c More than one triplet is changed, therefore more than one amino acid will also be changed. [2]
4 a To increase the amount of DNA from a small sample. [1]
  b 1 DNA is heated to 95 °C for 30 seconds. This separates the two strands by breaking the hydrogen bonds between the bases.
  2 DNA is rapidly cooled to 37 °C for 30 seconds. Primers bind to complementary strands of DNA.
  3 DNA is heated to 72 °C for two minutes. New strands of DNA are made.
  4 The cycle is repeated. [1 mark each]
  c They act as a template to start the process. [1]
5 a Gene is identified in donor DNA. Restriction enzymes cut out the DNA. The same enzymes cut open the host DNA in bacterial plasmid. The donor gene is added into the plasmid and the DNA joined by a ligase enzyme. The bacteria are cloned (allowed to multiply) to copy gene. [5]

  b To identify which bacteria have taken up the donor gene. [1]
6 a The CFTR gene is faulty so it cannot transfer Cl⁻ ions out of cell. [1]
  Water enters the cell by osmosis. [1]
  Secretions outside the cell thicken ie thick mucus forms. [1]
  b Viral DNA is destroyed and new CFTR gene is inserted into the virus. The modified virus is sprayed onto the patient's lungs. Virus injects DNA into epithelial cells so that they now code for normal CFTR protein. [3]
  c The enzyme Alpha 1 antitrypsin is made by human genes and helps protect lungs against damage from infection. Human DNA is inserted into the zygote of a sheep. The sheep produces the enzyme which can be given to patients. [3]
7 Answers will vary but see page 65 for ideas. Do not say simply that you are in favour of or against genetic engineering. You must give reasons for your opinion. [2]

Human health and disease (pp. 68-79)

1 a infection, auto-immune, diet-related, mental, environmentally caused, genetic [1 mark each]
  b better hygiene, eradication of TB in cattle, vaccination, X-ray screening [1 mark each]
2 a ii, iii, i [2]
  b It stays dormant for many years so the virus is passed on. [1]
  c The virus gets inside T-lymphocytes, safe from the immune system. [1]
  d intimate sex, sharing needles, infected blood products [1 mark each]
3 a It lives in blood and needs a mosquito vector to be passed on. [1]
  b Yes (no marks) because the malarial parasite can be seen. [1]
  c prophylactic drug treatment; insecticides; draining swamps; putting fish in swamps [1 mark each]
4 ear wax stops bacteria infecting ear; cilia in lungs trap and remove bacteria; tough outer layer of skin provides barrier; 'friendly' bacteria on skin prevent pathogenic bacteria multiplying; hydrochloric acid in stomach kills bacteria; blood clots to seal wounds; lysozyme in tears destroys bacteria entering the eyes. [5]
5 a Cells on left produce natural immunity; cells on right adaptive immunity. [2]
  b Natural immunity is non-specific; does not adapt; does not provide memory cells. [1 mark each]
6 a It produces plasma cells, which produce antibodies. [2]
  b The body can react quickly next time the disease is encountered. [2]
  c The cold virus mutates and changes antigens, so existing antibodies are no longer effective. [2]
  d Helper cells help B-cells produce antibodies; killer cells kill infected human cells; suppressor cells switch the immune system off. [1 mark each]

  e They make it easier for other cells in the immune system to deal with the bacteria. [2]
  f Any one of: rupture cell membranes, clump bacteria, neutralise toxins. [1]

Ecology and the environment (pp. 82-91)

1 a water, temperature, light, atmosphere, edaphic factors, e.g. soil [5]
  b a stable ecological unit, e.g. a wood or lake [1]
  c biosphere, biome, ecosystem, community [3]
2 a 1 chlorophyll 2 2H₂O 3 6-carbon sugar 4 CO₂ 5 NADPH₂ [5]
  b thylakoid membranes [1]
  c stroma [1]
3 a All the nutrients would be used up by the crop. [1]
  b Clover contains nitrogen-fixing bacteria which produce nitrates and these are ploughed into soil. [2]
  c It means they can use land all of the time and grow the crops they choose which is more cost effective. [2]
  d They contain nitrogen which is needed to produce protein. [2]
  e denitrifying bacteria; leaching / run-off [2]
4 i Can be used to attract insect pests which are then destroyed. [1]
  ii The females mate with them but produce no offspring, thus reducing their numbers. [1]
  iii Predators can be used to reduce the numbers of pests, e.g. *Encarsia* wasps laying eggs in whitefly larvae. [1]
5 a The farmer can spray wheat and kill the weeds, thus getting a weed-free crop. [2]
  b The farmer would have to buy wheat seeds and weed killer from same company. [2]
6 Fertilisers cause algae to grow; the algae block out sunlight; plants die and rot; bacteria use up oxygen; fish die. [5]
7 a It shows the population has few breeding adults to maintain the population. [1]
  b Very large nets are catching all the fish in an area. [1]
  c legislation; international agreements; fish farming [3]

Index